ILO Codes of Practice

Safety and health in building and civil engineering work

International Labour Office   Geneva 1972

ILO publications can be obtained through major booksellers or ILO local offices in many countries, or direct from ILO Publications, International Labour Office, CH-1211 Geneva 22, Switzerland. The catalogue and list of booksellers and local offices will be sent free of charge from the above address.

Printed by ATAR S.A. - Geneva (Switzerland)

Preface

This Code of Practice has been prepared in response to a wish expressed by the Building, Civil Engineering and Public Works Committee of the International Labour Organisation at its Seventh Session (Geneva, May 1964). The Committee—which is composed of representatives of governments, and of employers and workers in the construction industry—stated its belief that the Safety Provision (Building) Convention (No. 62) and the Safety Provision (Building) Recommendation (No. 53) do not take adequate account of new methods of construction which have come into use since these instruments were adopted by the International Labour Conference in 1937. To remedy this situation, the Committee, in its Resolution (No. 69) concerning action by the Office relating to technological developments and to safety in the construction industry [1], unanimously called for the preparation and publication of a Code of Practice on Safety in Building and Civil Engineering. In the light of a wish which the Committee expressed [2] on the same occasion for action by the Office relating to occupational health in the construction industry, it was decided that the present Code of Practice should also incorporate a number of provisions relating to occupational health in this industry.

The first draft of the present Code was prepared by the Office and was submitted, in 1968, for comments and observations to the 36 members of the ILO Panel of Consultants on Occupational Safety and Health in Building, Civil Engineering and Public Works. This Panel—which was set up pursuant to another wish expressed by the ILO Building, Civil Engineering and Public Works Committee in its above-mentioned Resolution (No. 69) concerning action by the Office relating to technological developments and to safety in the construction industry—is composed

---

[1] ILO: *Official Bulletin*, Vol. XLVII, No. 3, July 1964, p. 232.
[2] Ibid., p. 291.

of consultants specialising in the various aspects of the subject, and includes persons reflecting the views and experience of governments, as well as of employers' organisations and trade unions from this industry. The observations and comments of the members of the Panel were subsequently submitted for study to a small group of five consultants with a view to the preparation of the present consolidated text. At the invitation of the Office, the following organisations collaborated in this work: the Organisation for Economic Cooperation and Development; the Council of Europe; the International Organisation for Standardisation, the International Electrotechnical Commission, the European Mechanical Handling Federation, and the European Committee for Construction Equipment.

This document was approved for publication by the Governing Body of the ILO at its 180th Session (May-June 1970).

Contents

Safety and health in building and civil engineering

Introduction

The practical recommendations of this Code of Practice are intended for the use of all those, both in the public and in the private sectors, who have responsibility for safety and health in the building, civil engineering and public works industries. The Code is not intended to replace national laws or regulations or accepted standards. It has been drawn up with the object of providing guidance to those who may be engaged in the framing of provisions of this kind and, in particular, governmental or other public authorities, committees in civil engineering or public works establishments, and safety committees or management in related enterprises.

Local circumstances and technical possibilities will determine how far it is practicable to follow its provisions. Furthermore, these provisions should be read in the context of conditions in the country proposing to use this information. In this regard, the needs of the developing countries have also been taken into consideration.

Building and civil engineering is an extremely comprehensive subject and it may be divided into four main parts: work above ground, work in open excavations, underground work and underwater work. These parts can be subdivided in their turn. For instance, work above ground can be divided into building construction, public works construction and demolition. Detailed regulations have been drawn up under all these headings and in many different countries, and it is, therefore, not surprising that the present document is voluminous.

Many of the provisions are common to all industries (for instance, some of those concerning lifting appliances and gear, machines, vehicles, welding, painting and handling materials) but they have been included in the Code of Practice because it was felt that it should be as complete as practicable. The main exception is maintenance and repair shops, which may be assimilated to factories and thus come within the scope of the *Model*

Code of Safety Regulations for Industrial Establishments, which was first issued by the International Labour Office in 1949. Some countries include quarrying in their construction regulations, but for the purposes of this Code it is considered that quarrying is an extractive rather than a construction industry.

It is stressed that considerable thought was given to the differences which exist in building and civil engineering practices throughout the world, and the need both to improve poor practices and to establish good practices where none exist today. Special attention was also given to practical considerations relating to the establishment of certain recommended procedures. For example, in some areas adequate medical facilities are not available to the general population, and where this is so, it may be difficult to create satisfactory medical facilities on construction projects. Nevertheless, it is stressed that the provisions of the present Code of Practice represent the minimum that should be observed.

1. General provisions

1.1. Definitions

In this Code of Practice:

Adequate or *suitable* are used to describe qualitatively or quantitatively the means or method used to protect the worker.

Bearer see *putlog*.

Boatswain's chair means a seat for a workman which is suspended by a cable or rope.

Brace means a structural member that holds one point in a fixed position with respect to another point. *Bracing* is a system of structural members designed to prevent distortion of a structure.

Bracket scaffold means a scaffold composed of a platform supported on triangular braced brackets secured to the side of the building.

By hand as applied to work means that the work is done without the help of a mechanised tool.

Competent authority means a minister, government department, or other public authority having power to issue regulations, orders or other instructions having the force of law.

Competent person means a person who by reason of training or experience, or both, is competent to perform the task or function, or assume the responsibility, in question and is authorised to perform such task or function.

Contact voltage is that part of a fault voltage or voltage to earth that can make contact with a person.

Danger means danger of accident or injury to health.

Dust-tight or *dust-proof* applied to a machine or a device means that the machine or device is so constructed that dust of a specified fineness or nature cannot enter it or escape from it.

Earthed or *grounded* means connected to the general mass of earth in such a manner as will ensure at all times an immediate discharge of electrical energy without danger.

Grounded see *earthed*.

Guard-rail means an adequately secured rail erected along the exposed edges of scaffolds, floor openings, wall openings, runways, stairways, etc. to prevent persons from falling.

Hazard means danger or potential danger.

Inspection means an examination by which a competent person ascertains whether any place, working condition, construction, etc. presents any danger to workers.

Ladder jack scaffold means a scaffold, the platform of which is supported by jacks (brackets) attached to ladders.

Ladder scaffold means a scaffold, the platform of which is supported on ladders.

Ledger means a scaffold member which extends horizontally from post to post, at right angles to the putlogs, supports the putlogs, forms a tie between the posts, and becomes a part of the scaffold bracing. Ledgers which do not support putlogs are also called *stringers*.

Lifting appliance means a crane, hoist, derrick, winch, gin pole, sheer legs, jack, pulley block or other equipment for lifting materials, objects or, in some cases, persons.

Lifting gear means ropes, chains, hooks, slings and other accessories of a "lifting appliance".

Means of egress means passageways, corridors, stairs, platforms, ladders and any other means to be used by persons for normally leaving the workplace or for escaping in case of danger.

Outrigger scaffold means a scaffold, the platform of which is supported by outriggers or thrustouts projecting from the wall of the building, the inner ends of which are secured inside the building.

Pole scaffold may be a single pole scaffold or an independent pole scaffold. *Single pole scaffold* means a platform resting on putlogs or crossbeams, the outer ends of which are supported on ledgers secured to a single row of posts or uprights and the inner ends on a wall or in holes in a wall. *Independent pole scaffold* means a scaffold supported from the base by a double row of uprights, independent of support from the walls and constructed of uprights, ledgers, horizontal platform bearers, and diagonal bracing. An independent pole scaffold may also be referred to as a built-up scaffold.

Putlog or *bearer* means a scaffold member upon which the platform rests. In a single pole scaffold the outer end of the putlog rests on a ledger and the inner end rests in the wall. In an independent pole scaffold each end of the putlog rests on a ledger. In an independent pole scaffold a putlog is known as a bearer.

Safety testing means the action or process by which the properties of a substance, material, machine, etc. and the conditions obtaining in any workplace are examined with a view to determining their ability to satisfy certain prescribed standards of safety.

Scaffold means any temporary structure supporting one or more platforms used for supporting workmen or materials in the course of any type of construction work, including maintenance and demolition.

Sound construction means construction conforming to any relevant standards issued by a national standardising institution or other body recognised by the competent authority, or to generally accepted international engineering practices or other technical standards.

Sound material means material of a quality conforming to any relevant standards issued by a national standardising institution or other body recognised by the competent authority or to generally accepted international engineering practices or other technical standards.

5

Square scaffold means a scaffold, the platform of which is composed of planks supported on built-up squares secured to one other by diagonal bracing. *Square* means a framed structure built up of vertical and horizontal members and braces, which, when used in pairs and set up and braced longitudinally, provides a support for a working platform.

Stringer see *ledger*.

Sufficient means suitable in quantity to prevent danger.

Suspended scaffold means a scaffold supported at two or more points from overhead outriggers which are anchored to the building. It is equipped with a hoisting drum or machine so the platform can be raised or lowered.

Toe-board means a barrier placed along the edge of a scaffold platform, runway, etc. and secured there to guard against the slipping of persons or the falling of material.

Trestle scaffold means a scaffold, the platform of which is supported by trestles.

Voltage Safety extra-low voltage: see Chapter 17.

Window jack scaffold means a scaffold, the platform of which is supported by a jack or projection through a window opening.

1.2. General duties of employers

1.2.1. Employers should so provide and maintain buildings, plant, equipment and workplaces, and should so organise work, as to protect workers as far as practicable against risks of accidents and injuries to health.

1.2.2. When acquiring machines, appliances, vehicles or other equipment, employers should ensure that they conform to any official safety regulations applying to them or, if there are none, that they are so designed or protected that they can be operated safely.

1.2.3. Employers should provide such supervision as will ensure that as far as practicable workers perform their work in the best conditions of safety and health.

1.2.4. Work that is done jointly by a number of persons and requires mutual understanding and co-operation for the avoidance of risks should be specially supervised by a competent person.

1.2.5. Employers should only assign workers to employment for which they are suited by their age, sex, physique, state of health and skill.

1.2.6. Employers should not assign workers with physical or mental infirmities such as deafness, giddiness, bad sight and epilepsy to employment on which they could endanger themselves or others.

1.2.7. Employers should satisfy themselves that all workers are properly instructed in the hazards of their respective occupations and the precautions necessary to avoid accidents and injuries to health, and in particular that young workers, newly engaged workers, illiterate and foreign workers are properly instructed concerning hazards and precautions and are adequately supervised.

1.2.8. Employers should provide the workers with copies, extracts or summaries of national or local regulations and, whenever appropriate, instructions and notices relating to protection against accidents and injuries to health, or post such texts up in prominent positions at suitable places.

1.2.9. The regulations, instructions and notices should as far as practicable be in the languages of the workers employed.

1.2.10. Texts posted up should be undamageable or protected against damage from the weather, etc.

1.2.11. If practicable, separate safety rules should be drawn up for each type of occupation involved in a project.

1.2.12. Employers should ensure regular safety inspections by a competent person at suitable intervals of all buildings, equipment, workplaces and operations.

1.2.13. Employers should not permit buildings, equipment or workplaces in which a dangerous defect has been found to be used until the defect has been remedied.

1.2.14. If necessary, to prevent danger, employers should establish a checking system by which it can be ascertained whether all the members of a shift, including operators of mobile equipment, have returned to the camp or base at the close of work.

1.3. General duties of architects, engineers, designers

1.3.1. At the planning stage of any proposed building or civil engineering work, consideration should be given, by those responsible for the design, to the safety of the workers who will subsequently be employed in the erection of such structures.

1.3.2. Care should be exercised by architects, including inside architects, engineers and other professional persons, not to include anything in the design which would necessitate the use of unwarrantably dangerous structural procedures and undue hazards, which could be avoided by design modifications.

1.3.3. It is also of the greatest importance that structural designers should take into account the safety problems associated with the subsequent maintenance of structures where this would involve special hazards.

1.3.4. Facilities should be included in the design for such work to be performed with the minimum of risk.[1]

1.4. General duties of workers

1.4.1. Within the limits of their responsibilities, workers should do everything in their power to maintain their and their workmates' health and safety.

---

[1] It is, for example, essential that there should be included in the design of a tall building facilities to enable work such as the outside cleaning of windows to be carried out by safe methods.

1.4.2. Before beginning work workers should examine their workplaces and the equipment that they are to use, and should forthwith report to their foreman or other competent superior any dangerous defect that they may discover in them.

1.4.3. Workers should make proper use of all safeguards, safety devices and other appliances furnished for their protection or the protection of others.

1.4.4. Except in an emergency, no worker, unless duly authorised, should interfere with, remove, alter, or displace any safety device or other appliance furnished for his protection or the protection of others, or interfere with any method or process adopted with a view to avoiding accidents and injuries to health.

1.4.5. Workers should not interfere with equipment such as machines and appliances that they have not been duly authorised to operate, maintain or use.

1.4.6. Workers should not sleep or rest in dangerous places such as scaffolds, railway tracks, garages, or in the vicinity of fires, dangerous or toxic substances, running machines or vehicles and heavy equipment.

1.4.7. Workers should make themselves acquainted with and obey all safety and health instructions pertaining to their work.

1.4.8. Workers should refrain from careless or reckless practices or actions that are likely to result in accidents or injuries to health to themselves or others.

1.4.9. Workers should wear protective equipment and clothing suited to their duties and to the weather conditions.

1.4.10. Workers should practise good housekeeping. (See section 2.3.)

1.5. Obligations of manufacturers and dealers

1.5.1. In order to prevent dangerous equipment from reaching users and to ensure that the necessary precautions are taken by users, manufacturers and dealers should ensure that:

(a) equipment such as machines, appliances and vehicles used in the construction industry complies with national or other official safety laws and regulations, and standards applicable to its design and construction;

(b) equipment not covered by national or other official laws and regulations or standards is so designed and constructed as to be as safe as practicable; and

(c) equipment is accompanied by printed matter giving the necessary instructions for its proper testing, use and maintenance and drawing attention to possible hazards.

1.5.2. Manufacturers and vendors of flammable liquids, explosives, toxic, corrosive and other dangerous substances should furnish users with adequate instructions for their safe use.

1.6. Employment of young persons under 18

1.6.1. No person under 16 years of age should be employed in the construction industry except as permitted by the competent authority and in accordance with conditions prescribed by that authority.

1.6.2. No person under 18 years of age should be employed on work that is particularly dangerous, or is liable to affect the safety or health of considerable numbers of workers, or requires mature judgment for its safe performance, such as the operation of power-driven machinery, cranes, driving tractors, handling flammable liquids in bulk, work with explosives, the operation of steam boilers, and work with toxic or corrosive substances.

1.6.3. Young persons under 18 years of age should not be employed or work during the night in the construction industry except for vocational training purposes in the cases and under conditions to be specified by the competent authority.[1]

---

[1] See also ILO Convention No. 90, the Night Work of Young Persons (Industry) Convention (Revised), 1948.

1.6.4.   The restrictions of this section apply to all building and civil engineering work, including constructional, repair, maintenance, alteration and demolition work.

1.7. Employment of women

1.7.1.   Women should be employed in accordance with the provisions of national laws and regulations, or if there are none, with provisions that should be enacted, concerning:

(a) work before and after childbirth;
(b) night work;
(c) lifting, carrying and moving loads;
(d) handling dangerous substances; and
(e) performing dangerous or unhealthy operations.

1.8. Signalling

Signal code

1.8.1.   Employers should establish a system of signalling for all operations in which signals are required to prevent danger.

1.8.2.   As far as practicable, a uniform signalling system should be adopted for all construction sites in the same country.

1.8.3.   The code of signals should be posted up at suitable places and also made available in the form of a handbook.

1.8.4.   In order to avoid danger, employers should take adequate steps to ensure that workers are familiar with all signals that they should know.

1.8.5.   Each signal should only have one meaning.

Signallers

1.8.6.   Signals should only be given by reliable, competent persons duly authorised to give signals.

1.8.7.   No operation should be governed by more than one signaller in charge of the operations; this does not exclude one or more assistant signallers as may be necessary to transmit signals to the person operating the machinery.

1.8.8. Signallers should have no other duties when signalling.

1.8.9. Signallers should be prepared to give the stop signal at any moment while signalling; a "stop" signal may be given by anyone in case of an emergency.

Operating requirements

1.8.10. No operation for which a signal is provided in the code should be carried out until that signal has been given.

1.8.11. No signal not provided for in the code should be given or obeyed.

1.8.12. Hand signals should only be given when all persons for whom they are intended can easily see them.

1.8.13. Audible signals should be clearly audible to all persons whom they are intended to protect and who might be endangered if they did not hear them.

1.8.14. Any signal that is not properly understood should be treated as a stop signal.

1.8.15. No signal for the movement of equipment should be given until the signaller has satisfied himself that no person in the area for which he is responsible will be endangered by the movement.

Signalling equipment

1.8.16. The signaller's workplace should be:

(a) safe from moving equipment, falling objects and other hazards;
(b) such that the signaller has an unobstructed view of the operations that he is directing; and
(c) such that the persons concerned can easily hear or see the signals.

1.8.17. Signalling equipment should be efficient, properly installed, regularly tested and kept in good working order.

1.8.18. Only competent persons should repair, alter or adjust signalling devices.

1.8.19. Radio-frequency signalling equipment should have the frequency conspicuously marked on both the transmitter and the receiver.

1.8.20. Radio-frequency signalling equipment should not affect or be affected by any other signalling equipment in the neighbourhood.

1.8.21. In case of electrical storms which can affect the transmission, no radio signals whose misunderstanding might lead to an accident should be given.

2. Workplaces and equipment

2.1. Means of access and egress

2.1.1.   As far as is reasonably practicable, adequate and safe means of access and egress should be provided for all workplaces.

2.1.2.   Means of access and egress should be maintained in a safe condition.

2.1.3.   Where special safe means of access to or egress from workplaces are provided, workers should always use them for going to and from the workplaces.

2.2. Heating, lighting, ventilation

Heating

2.2.1.   Workplaces should be adequately heated wherever necessary and practicable. If it is not practicable, provisions should be made to allow workers to warm themselves at appropriate places and intervals during their work.

2.2.2.   Heating installations should comply with the requirements of paragraphs 2.4.13 to 2.4.23.

Lighting

2.2.3.   All practicable measures should be taken to prevent steam, smoke, fumes, etc. from obscuring any workplace or equipment at which any person is employed.

2.2.4.   Where natural lighting is not adequate to prevent danger, adequate and suitable artificial lighting should be provided at all workplaces and their approaches, including passageways.

2.2.5.   Artificial lighting should not cause any danger, including that of producing glare or disturbing shadows.

2.2.6.   Lamps should be protected by suitable guards where necessary to prevent danger if a lamp breaks.

Ventilation

2.2.7. In enclosed workplaces, suitable provision should be made to ensure an adequate circulation of fresh air.

2.2.8. Where necessary to prevent danger to health from air contamination by dust from the grinding, cleaning, spraying or manipulation of materials or objects, or from dangerous gases or from any other cause, arrangements should be made for the removal or dilution to safe limits of contaminants by means of ventilation. Particular attention should be given to the control cabins of cranes, driving cabs of lorries and similar confined workplaces.

2.2.9. If it is not technically possible to eliminate dust or noxious or harmful fumes or gases sufficiently to prevent injury to health, the workers should be provided with respiratory protective equipment, complying with the requirements of paragraphs 36.1.38 to 36.1.45.

2.3. Housekeeping

2.3.1. Loose materials which are not required for use should not be placed or left so as dangerously to obstruct workplaces and passageways.

2.3.2. All projecting nails should be removed or bent over to prevent injury.

2.3.3. Equipment, tools and small objects should not be left lying about where they could cause an accident either by falling or causing a person to trip.

2.3.4. Scrap, waste and rubbish should not be allowed to accumulate on the site.

2.3.5. Workplaces and passageways that are slippery owing to ice, snow, oil or other causes should be cleaned up or strewn with sand, sawdust, ash or the like.

2.3.6. Portable equipment should be returned after use to its designated storage place.

2.4. Fire protection

Fire-extinguishing equipment

2.4.1. Places where workers are employed should, if necessary to prevent danger, be provided as far as practicable with:

(a) suitable and sufficient fire-extinguishing equipment; and

(b) an adequate water supply at ample pressure.

2.4.2. All supervisors and a sufficient number of workers should be trained in the use of fire-extinguishing equipment.

2.4.3. Persons trained to use the fire-extinguishing equipment should be readily available during all working periods.

2.4.4. Fire-extinguishing equipment should be inspected at suitable intervals by a competent person and properly maintained.

2.4.5. Access to fire-extinguishing equipment such as hydrants, portable extinguishers and connections for hoses should be kept clear at all times.

2.4.6. Fire-extinguishing equipment should be easily visible.

2.4.7. At least one adequate fire extinguisher should be provided:

(a) in every building where combustible materials are stored;

(b) at places where any welding and flame-cutting operations are carried out; and

(c) on each floor of a building that is being constructed or altered and where there is combustible material.

2.4.8. The necessary number of suitable dry chemical extinguishers should be provided:

(a) where flammable liquids are stored or handled;

(b) where oil- or gas-fired heating equipment is used;

(c) where a tar or asphalt kettle is used; and

(d) where there is a danger of electrical fires.

2.4.9. Fire-extinguishing equipment should be adequately protected against mechanical damage.

2.4.10. In cold weather, extinguishers should be protected against freezing.

2.4.11. Extinguishers containing chlorinated hydrocarbons or carbon tetrachloride should not be used indoors or in confined spaces.

2.4.12. Where a standpipe is to be installed in a building it should:

(a) be installed progressively as far as practicable as the construction proceeds;

(b) be provided with a valve at each hose outlet;

(c) be provided at each hose outlet with a nozzle and sufficient suitable hose; and

(d) have a suitable connection for the public fire department.

Heating appliances

2.4.13. Open-flame heating appliances such as braziers should only be used in places where there is adequate general ventilation.

2.4.14. No open-flame heating appliance should be placed in the means of egress.

2.4.15. Combustion appliances such as stoves, salamanders and braziers should not be placed on wooden floors or other combustible bases, but on non-combustible bases extending a safe distance beyond the stove or brazier on all sides.

2.4.16. Combustion heating appliances used indoors should be equipped with an adequate device to discharge the combustion gases into the outside air.

2.4.17. Combustion heating appliances used indoors should be kept at a safe distance from combustible structures and material.

2.4.18. Tarpaulins, canvas sheets and the like in the vicinity of any appliance likely to ignite them should be secured so that they cannot blow on to or come in contact with the heating surface.

2.4.19. Braziers should preferably not use bituminous coal.

2.4.20. Flues and chimneys, even for temporary installations, should be adequately insulated when passing through combustible walls, ceilings, roofs and the like.

2.4.21. Heating systems for stores of flammable or combustible materials should not have any open flame or exposed incandescent part.

2.4.22. No part of heating systems for stores of flammable or combustible materials should be in dangerous proximity to such materials.

2.4.23. Fires should not be started with flammable liquids such as oil and petrol (gasoline).

Combustible materials

2.4.24. Combustible material such as sawdust, greasy rags and scrap wood should not be allowed to accumulate at workplaces.

2.4.25. Oily clothing should not be left in confined spaces.

2.4.26. Unslaked lime should be kept dry.

2.4.27. In buildings, oily waste should be kept in metal containers with self-closing lids.

2.4.28. There should be no smoking and no open flame or incandescent material in dangerous proximity to flammable or combustible material.

Flammable liquids

2.4.29. Flammable liquids should be stored, transported, handled and used in conformity with the requirements of section 21.2.

2.4.30. No fuel for a heating device should be stored in a building or structure except in a fire-resistant container constructed for the purpose.

2.4.31. No fuel should be stored in means of egress.

Inspection, supervision

2.4.32. Regular inspection should be made of places where there are fire risks. These include, namely, the vicinity of heating appliances, electrical installations and conductors, stores of flammable liquids and combustible materials, welding operations, internal combustion engines and, if necessary to prevent danger, hot roofing.

2.4.33. When necessary to prevent danger from fire, a competent person should be on duty at construction sites out of working hours.

Electrical installations

2.4.34. Electrical installations should comply with the relevant provisions of Chapter 17.

Notices

2.4.35. Notices should be posted at conspicuous places indicating:

(a) the nearest fire alarm; and
(b) the telephone number and address of the nearest fire brigade.

2.5. Protection against falls of objects and collapse of structures and materials

2.5.1. Where necessary to prevent danger, overhead screens of adequate strength and dimension should be provided or other effective precautions should be taken to prevent workers from being struck by falling objects.

2.5.2. Materials and objects such as, for example, scaffolding materials, waste materials and tools should not be thrown up, or down from heights, if they are liable to cause injury to any person.

2.5.3. If materials and objects cannot be safely lowered from heights, adequate precautions such as the provision of fencing, lookout men or barriers should be taken to protect any person who might be injured by them.

2.5.4. Workers should not enter silos, bunkers, chutes and the like unless:

(a) duly authorised to do so;
(b) the discharge opening is closed and secured against opening;

(c) the worker wears a safety belt with a lifeline securely attached to a fixed object and complying with the requirements of Chapter 36;

(d) the worker is under constant surveillance by another person; and

(e) as far as silos are concerned, the requirements of Chapter 20 are complied with.

2.5.5. Where necessary to prevent danger, guys, stays or supports should be used or other effective precautions should be taken to prevent the collapse of structures or parts of structures that are being erected, maintained, repaired or demolished.

2.6. Protection against falls of persons

Guard-rails and toe-boards (railings)

2.6.1. All guard-rails and toe-boards for the fencing of floor openings, wall openings, gangways, elevated workplaces and other places to prevent falls of persons should:

(a) be of sound material and good construction and possess adequate strength;

(b) be between 1 m (3 ft 3 in) and 1.15 m (3 ft 9 in) above platform level as regards guard-rails;

(c) consist of (i) two rails (two ropes or chains may be used if they are sufficiently taut); (ii) supporting stanchions; and, (iii) if necessary to prevent persons slipping or objects falling, a toe-board.

2.6.2. Intermediate rails, ropes or chains should be midway between the top of the toe-board and lower edge of the top rail.

2.6.3. A sufficient number of stanchions or standard poles or uprights should be maintained to ensure sufficient stability and resistance.

2.6.4. Toe-boards should be at least 15 cm (6 in) high, and securely fastened.

2.6.5. Guard-rails and toe-boards should be free from sharp edges, and be maintained in good repair.

Floor openings

2.6.6.   Floor openings through which persons could fall should be guarded:

(a) by covers complying with the requirements of paragraphs 2.6.8 to 2.6.12;

(b) by guard-rails and toe-boards on all exposed sides complying with the requirements of paragraphs 2.6.1 to 2.6.5; or

(c) by other effective means.

2.6.7.   If the means of protection is removed to allow the passage of persons or goods or for other purposes, it should be replaced as soon as practicable.

2.6.8.   Covers for floor openings should be safe to walk on, and, if necessary, safe to drive vehicles on.

2.6.9.   Covers for floor openings should be secured by hinges, grooves, stops or other effective means against sliding, falling down or lifting out, or any other inadvertent displacement.

2.6.10.   Covers for floor openings should not constitute any hindrance to traffic and, as far as practicable, be flush with the floor.

2.6.11.   If covers are constituted as grids, the bars should be spaced not more than 5 cm (2 in) apart.

2.6.12.   Covers of sack and similar elevators should close automatically after the passage of the load.

Wall openings

2.6.13.   Wall openings less than 1 m (3 ft 3 in) from the floor and measuring at least 75 cm (2 ft 6 in) vertically and 45 cm (1 ft 6 in) horizontally from which there is a drop of more than 1.5 m (5 ft) should be protected to a height of at least 1 m (3 ft 3 in) by guard-rails and toe-boards complying with the requirements of paragraphs 2.6.1 and 2.6.2 of this section, or other effective means.

2.6.14.   Narrower openings should be protected by a toe-board if their lower edge is less than 15 cm (6 in) from the floor.

2.6.15. If the protection of openings is removable:

(a) an adequate handgrip should be fixed on each side; or

(b) an adequate bar should be placed across the opening to prevent falling.

Elevated workplaces

2.6.16. Elevated workplaces more than 2 m (6 ft 6 in) above the floor or ground should be protected on all open sides by guardrails and toe-boards complying with the requirements of paragraphs 2.6.1 and 2.6.2.

2.6.17. Elevated workplaces should be provided with safe means of access and egress such as stairs complying with the requirements of section 4.7, ramps complying with the requirements of section 3.3, or ladders complying with the relevant requirements of Chapter 4.

2.6.18. Where necessary to prevent danger, persons employed at elevated workplaces and other workplaces from which they might fall more than 2 m (6 ft 6 in) should be protected by means of adequate catch nets, sheets or platforms, or be secured by safety belts with the lifeline properly attached.

Protection against falls into water; rescue arrangements

2.6.19. Where workers are in danger of falling into water and drowning they should wear a life vest; adequate rescue arrangements, such as the provision of a suitable manned boat and ring buoys, should be made and constantly maintained for the duration of the danger, in conformity with requirements of Chapter 19.

2.7. Noise and vibrations

2.7.1. Noise and vibrations likely to have harmful effects on workers should be reduced to practicable levels.

2.7.2. Noise should be reduced to the appropriate values, which should be fixed by competent authorities.

2.7.3. If noise cannot be rendered harmless, workers should be provided with suitable ear protectors.

2.8. Protection against unauthorised persons

2.8.1. Construction sites in built-up areas and alongside main traffic routes should be barricaded.

2.8.2. Unauthorised persons should not be allowed access to construction sites unless accompanied by a competent person and provided with the protective equipment required for sites concerned.

2.9. Structures and equipment

Construction

2.9.1. Structures (for example scaffolding, platforms, gangways and towers) and equipment (for example machines, lifting appliances, pressure plant and vehicles, used on a construction site) should:

(a) be of sound material and of good quality;

(b) be free from patent defect; and

(c) be properly constructed in accordance with sound engineering principles.

2.9.2. Structures and equipment should be strong enough to withstand safely the loads and stresses to which they will be subjected.

2.9.3. Metal parts of structures and equipment should:

(a) not have been weakened by cracks, rust or corrosives; and

(b) if necessary to prevent danger, be given a protective coating.

2.9.4. Wooden parts of structures and equipment such as scaffolding, shuttering and ladders, should:

(a) have the bark completely removed; and

(b) not be painted so as to conceal defects.

2.9.5. Used wood should be freed from nails, iron straps, etc. before it is used again.

Inspection, testing, maintenance

2.9.6.   Structures and equipment should be inspected before being taken into use and at suitable intervals by a competent person.

2.9.7.   Before being taken into use, and at suitable intervals, structures and equipment that might cause serious accidents if defective, for instance pressure plant, lifting appliances, and scaffolding, should be inspected and/or tested by a competent person in accordance with requirements which should be laid down by the competent authority.

2.9.8.   Structures and equipment should be constantly maintained in a safe condition.

2.9.9.   Structures and equipment should be specially inspected by a competent person:

(a) after any breakdown or any damage liable to cause danger;
(b) after any accident in which they have been involved;
(c) after any substantial alteration; and
(d) after any dismantling, transfer to another site, or re-erection.

2.9.10.   Equipment such as scaffolding, shuttering and bracing and tower cranes should be inspected:

(a) after any lengthy idle period;
(b) after a high wind or heavy rain;
(c) after heavy vibration due to earth tremors, blasting, or any other cause.

2.9.11.   Structures and equipment found to be dangerously defective should be taken out of use and not used again until they have been made safe.

2.9.12.   If necessary to prevent danger, defective structures and equipment that have been repaired should not be used again until they have been inspected and tested by a competent person or, in the case of structures and equipment referred to in paragraph 2.9.7, by a competent person in accordance with the requirements of the competent authority.

2.9.13. Particulars and results of all inspections of structures and equipment should be entered in a special register.

Use

2.9.14. Structures and equipment should only be used for the purposes for which they were intended.

2.9.15. Equipment should only by operated by a competent person or persons.

3. Scaffolds

3.1. General provisions

3.1.1. Suitable and sufficient scaffolds should be provided for workers for all work that cannot safely be done at a height from a ladder or by other means.

3.1.2. A scaffold should only be constructed, taken down and substantially altered:

(a) under the direction of a competent and responsible person; and
(b) as far as practicable by competent persons.

Materials

3.1.3. Sufficient material should be provided for and used in the construction of scaffolds.

3.1.4. Timber used in the construction of scaffolds should be straight-grained, sound, and free from large knots, dry rot, worm holes and other dangerous defects.

3.1.5. No rope that has been in contact with acids or other corrosive substances or is defective should be used on scaffolds.

3.1.6. No fibre rope should be used on a scaffold erected at any place where such rope would be liable to be damaged.

3.1.7. Where necessary, boards and planks used for scaffolds should be protected against splitting.

3.1.8. Nails on scaffolds should be of adequate length and thickness.

3.1.9. Cast-iron nails should not be used on scaffolds.

3.1.10. Materials used in the construction of scaffolds should be stored under good conditions and apart from any material unsuitable for scaffolds.

3.1.11. Fastenings on wooden scaffolds should be steel bolts of adequate dimensions with washers and nuts, fibre rope lashings, nails, adequate clamps, or other means approved by the competent authority.

Construction

3.1.12. Scaffolds should be designed with a safety factor of four times their maximum load.

3.1.13. Pole, ladder and similar scaffolds should be provided with safe means of access such as stairs, ladders, or ramps.

3.1.14. Pole, ladder and similar scaffolds should be adequately braced.

3.1.15. Pole, ladder and similar scaffolds which are not independent should be rigidly connected to the building at suitable vertical and horizontal distances.

3.1.16. A scaffold should never extend above the highest anchorage to an extent which might endanger its stability and strength.

3.1.17. On independent scaffolds, sufficient putlogs and transoms should remain in position and securely fastened to the ledgers, uprights or standards, as the case may be, to ensure the stability of the scaffold until it is finally dismantled.

3.1.18. All structures and appliances used as supports for working platforms should be of sound construction, have a firm footing, and be adequately strutted and braced to make them stable.

3.1.19. Loose bricks, drain pipes, chimney pots or other unsuitable material should not be used for the construction or support of scaffolds.

3.1.20. When necessary to prevent danger from falling objects, scaffolds should be provided with adequate screens.

3.1.21. Nails should be driven full length, and not driven part way and then bent over.

3.1.22. No nail should be subjected to direct pull.

Inspection, maintenance

3.1.23. Every scaffold should, before use, be examined by a competent person to ensure more particularly:

(a) that it is in a stable condition;

(b) that the materials used in its construction are sound;

(c) that it is adequate for the purpose for which it is to be used; and

(d) that the required safeguards are in position.

3.1.24. Scaffolds should be inspected by a competent person:

(a) at least once a week; and

(b) after every spell of bad weather and every prolonged interruption in the work.

3.1.25. Scaffold parts should be inspected on each occasion before erection.

3.1.26. Every scaffold should be maintained in good and proper condition, and every part should be kept fixed or secured so that no part can be displaced in consequence of normal use.

3.1.27. No scaffold should be partly dismantled and left so that it is capable of being used, unless it continues to be safe for use.

Lifting appliances on scaffolds

3.1.28. When a lifting appliance is to be used on a scaffold:

(a) the parts of the scaffold should be carefully inspected and, if need be, adequately strengthened;

(b) any movement of the putlogs should be prevented; and

(c) if practicable, the uprights should be rigidly connected to a solid part of the building at the place where the lifting appliance is erected.

3.1.29. When the platform of the lifting appliance does not move in guides or when the load is liable to come into contact with the scaffold during hoisting or lowering, a vertical hoarding should be erected to the full height of the scaffold to prevent loads from being caught in it.

3.1.30.   Jibs or gallows for hoisting materials should not be attached to standards or extension poles.

3.1.31.   When no jib or gallows but only a rope pulley is used, the pulley should not be attached to a crossbeam unless the crossbeam:

(a) has sufficient strength and is fixed to at least two uprights or extensions in the way prescribed for ledgers; and
(b) does not at the same time serve as a ledger for the scaffold.

3.1.32.   If a lifting appliance or any part of one moves along a scaffold, adequate measures should be taken to prevent persons on the scaffold from being struck by the appliance or any part of it.

3.1.33.   When lifting appliances running on masts, in openwork metal towers, on inclines or otherwise are attached to or lean on scaffolds:

(a) the lifting installation should be erected on a firm level base;
(b) where necessary to prevent danger the scaffold should be additionally braced;
(c) unless it is an independent self-supporting structure, the lifting installation should be tied to the scaffold at suitable intervals;
(d) the travelway of the lifting appliance should extend to at least 2.5 m (8 ft) above the top landing;
(e) the bottom landing should be protected from falling objects, for example, by a roof; and
(f) the landings above ground level should be protected by an enclosure or a gate the height of which should not be less than 2 m (6 ft 6 in) above platform level to prevent persons coming into contact with moving parts.

Prefabricated frames

3.1.34.   Prefabricated frames for scaffolds should have adequate arrangements on both faces for fixing bracing, and if necessary to prevent danger.   They should also have guard-rails.

3.1.35.   Frames of different types should not be intermingled.

3.1.36.   Frames should be sufficiently strong and rigid to avoid distortion during transport, handling, etc.

3.1.37.   Where frames are superimposed vertically, adequate precautions should be taken to keep the legs in correct alignment.

3.1.38.   On independent scaffolding, adequate precautions should be taken to prevent vertical separation of frames.

Use of scaffolds

3.1.39.   In transferring heavy loads on or to a scaffold no sudden shock should be transmitted to the scaffold.

3.1.40.   When necessary to prevent danger, loads being hoisted on to scaffolds should be controlled by a hand rope (tag line) so that they cannot strike against the scaffold.

3.1.41.   The load on the scaffold should be evenly distributed as far as is practicable, and in any case should be so distributed as to avoid any dangerous disturbance of the equilibrium.

3.1.42.   During the use of a scaffold care should constantly be taken that it is not overloaded.

3.1.43.   Scaffolds should not be used for the storage of material except that required for immediate use.

3.1.44.   Workers should not be employed on outside scaffolds in a high wind.

3.1.45.   In order to prevent damage to scaffold materials, such materials should be properly lifted and lowered.

3.2. Working platforms

General provisions

3.2.1.   All scaffolds on which workers are employed should be provided with a sufficient number of working platforms.

3.2.2.   No part of a working platform should be supported by loose bricks, drain pipes, chimney pots or other loose or unsuitable material.

3.2.3. No working platform should be supported by an eaves gutter, a balcony or its coping, a lightning conductor or other unsuitable parts of a building.

3.2.4. No working platform should be used for working upon until its construction is completed and the necessary safeguards properly fixed.

3.2.5. Whenever practicable, a platform should extend at least 60 cm (2 ft) beyond the end of the wall of the building.

3.2.6. The width of the platform should be adequate having regard to the nature of the work, and should be such that at every part there is not less than 60 cm (2 ft) clear passage free from fixed obstacles and deposited material.

3.2.7. In no case should the width of a working platform be less than:

(a) 60 cm (2 ft) if the platform is used as a footing only and not for the deposit of any material;

(b) 80 cm (2 ft 8 in) if the platform is used for the deposit of material;

(c) 1.1 m (3 ft 8 in) if the platform is used for the support of any higher platform;

(d) 1.3 m (4 ft 4 in) if the platform is one upon which stone is dressed or roughly shaped;

(e) 1.5 m (5 ft) if the platform is used for the support of any higher platform and is one upon which stone is dressed or roughly shaped.

3.2.8. The maximum width of a platform supported on putlogs should as a rule not exceed 1.6 m (5 ft 4 in).

3.2.9. As far as practicable, a clear headroom of at least 1.8 m (6 ft) should be maintained on working platforms.

3.2.10. Every working platform should, if part of a pole scaffold, be at least 1 m (3 ft 3 in) below the top of the uprights.

3.2.11. Every working platform more than 2 m (6 ft 6 in) above the ground or floor should be closely boarded or planked.

3.2.12. Boards or planks which form part of a working platform or which are used as toe-boards should be of dimensions such as to afford adequate security having regard to the distance between the putlogs, but in no case should they:

(a) be of a thickness less than 2.5 cm (1 in); and

(b) be of a width less than 15 cm (6 in).

3.2.13. No board or plank which forms part of a working platform should project beyond its end support to a distance exceeding four times the thickness of the board or plank.

3.2.14. If practicable, boards or planks should not overlap one another; if not, precautions such as the provision of bevelled pieces should be taken to reduce the risk of tripping to a minimum and to facilitate the movement of wheelbarrows.

3.2.15. Planks used for flooring should be of uniform thickness.

3.2.16. Every board or plank which forms part of a working platform should rest on at least three supports, unless the distance between the putlogs and the thickness of the board or plank is such as to exclude all risk of tipping or undue sagging.

3.2.17. Platforms should be so constructed that the boards or planks cannot be displaced through normal use.

Guard-rails and toe-boards (railings)

3.2.18. Every part of a working platform or working place from which a person is liable to fall a distance exceeding 2 m (6 ft 6 in) should be provided with railings complying with the requirements of paragraphs 2.6.1 to 2.6.5.

3.2.19. Guard-rails, toe-boards and other safeguards used on a scaffold platform should be maintained in position, except for the time and to the extent required to allow the access of persons or the transport or shifting of materials.

3.2.20. Guard-rails and toe-boards used on a scaffold platform should be placed on the inside of the uprights, except where provision has been made in the design to prevent any outward movement.

Suspended platforms

3.2.21.   The platforms of suspended scaffolds should be provided with guard-rails and toe-boards on all sides, except that:

(a) on the side facing the wall the guard-rail need not be at a height of more than 70 cm (2 ft 4 in) if the work does not allow of a greater height;

(b) the guard-rails and toe-boards should not be compulsory on the side facing the wall if the workers sit on the platform to work, but in such case the platform should be provided with cables, ropes or chains affording the workers a firm handhold and capable of holding any worker who may slip.

3.2.22.   The space between the wall and the platform should be as small as practicable except where workers sit on the platform during their work, in which case it should not exceed 45 cm (1 ft 6 in).

Platform suspended from lifting equipment

3.2.23.   When a working platform is suspended from lifting equipment, if necessary to prevent danger the lifting equipment should have means for positively locking the supports so as to prevent inadvertent movement of the platform.

3.2.24.   If necessary to prevent danger, the lifting equipment operator should remain at the controls while the platform is in use.

3.2.25.   If the platform is supported from overhead:

(a) adequate precautions should be taken to prevent spinning; and

(b) the operation of the platform should be governed by signals in conformity with the requirements of section 1.8.

3.2.26.   Precautions should be taken to prevent the platform from tilting during raising and lowering.

3.2.27.   While the suspended platform is in use the lifting equipment should not be moved on any surface, and in any conditions, that would dangerously affect the stability of the platform.

3.2.28. Men working on suspended platforms should be provided with and wear safety belts which should be connected to lifelines independent of the platform or its suspension.

3.3. Gangways, ramps and runways

3.3.1. Gangways, ramps and runways should be so constructed and supported that they cannot tip, sag unduly or collapse under the maximum loads which they will have to carry.

3.3.2. Every gangway, ramp and runway any part of which is more than 2 m (6 ft 6 in) above the ground or floor should be:

(a) closely boarded or planked; and
(b) at least 60 cm (2 ft) wide.

3.3.3. When a gangway, ramp or runway is used for the passage of materials, there should be maintained a clear passageway which:

(a) is adequate in width for transport of materials without the removal of guard-rails and toe-boards; and
(b) is in any case of a width not less than 60 cm (2 ft).

3.3.4. The slope of any gangway, ramp or runway should not exceed 1 in 4.

3.3.5. When the slope renders additional foothold necessary there should be proper stepping laths which should:

(a) be placed at equal intervals appropriate to the gradient; and
(b) be the full width of the gangway, ramp or runway except for a 10 cm (4 in) gap to facilitate the movement of wheelbarrows.

3.3.6. Gangways, ramps and runways from which a person could fall more than 2 m (6 ft 6 in) should be provided with guard-rails complying with the requirements of paragraphs 2.6.1 to 2.6.5.

3.3.7. Ramps or runways built to provide access for vehicles on a construction site should:

(a) possess adequate strength and stability to withstand safely the maximum loads that they will have to carry;
(b) have a gradient and a width safe for the vehicles; and
(c) be provided with a substantial curb on each side.

3.3.8.   The minimum width inside the curbs should be 60 cm (2 ft) more than the width of the widest vehicle using the ramp or runway if there is no passing, or 60 cm (2 ft) more than twice this width if there is passing.

3.3.9.   Ramps and runways should be level transversely.

3.4. Wooden pole scaffolds

Uprights

3.4.1.   Pole uprights should be:

(a) vertical or slightly inclined towards the building; and
(b) placed sufficiently close together to secure the stability of the scaffolds.

3.4.2.   The diameters of poles should:

(a) be determined in accordance with expected maximum loading;
(b) be at least 8 cm (3½ in) at the level of the highest ledger.

3.4.3.   If pole uprights have to be extended:

(a) the upper and lower poles should overlap by at least 1.5 m (5 ft);
(b) the two poles should be securely wedged together and lashed with wire, wire rope, chains or the like; and
(c) the extension pole should rest on a ledger, putlog, or other adequate support.

3.4.4.   The stability of pole uprights should be secured:

(a) by letting the pole an adequate distance into the ground according to the nature of the soil; or
(b) by placing the pole on a suitable plank or other adequate sole-plate in such a manner as to prevent slipping; or
(c) in any other sufficient way.

3.4.5.   When two scaffolds meet at the corner of a building a pole upright should be placed at the corner on the outside of the scaffolds.

3.4.6. Where splices are necessary in round pole uprights:

(a) butt-jointed double poles should be used instead of single poles;

(b) the joins of the two poles should be as far as possible staggered; and

(c) the poles should be adequately lashed together at the base and at each butt joint.

3.4.7. When necessary the bases of poles should be provided with suitable protection against the impact of trucks or other moving equipment.

3.4.8. Sawn timber uprights should be butt jointed with adequate timber pieces or fish-plates secured on each side of the butt joint with bolts, washers and nuts.

Ledgers

3.4.9. Ledgers should be practically level and securely fastened to the uprights by bolts, dogs, ropes or other efficient means.

3.4.10. The ends of two consecutive ledgers at the same level should be securely joined together at an upright except when special devices are used which ensure equivalent strength.

3.4.11. Two consecutive ledgers should overlap at least 1 m (3 ft 3 in).

3.4.12. Unsupported ends of ledgers should be avoided. In any case no load should be placed on their ends.

3.4.13. The vertical distance between rows of ledgers should not exceed 4 m (13 ft).

3.4.14. Ledgers should extend over the whole length of the scaffold.

3.4.15. When necessary to prevent danger from heavy loading, ledgers should be adequately reinforced by bracing, cleats or other effective means.

3.4.16. All ledgers should be left in place to brace the scaffold until it is dismantled.

Putlogs

3.4.17.   Putlogs should be in one piece, straight and securely fastened to the ledgers.

3.4.18.   If ledgers are not used, the putlogs should be fastened to the uprights and supported by securely fastened cleats.

3.4.19.   If one end of a putlog cannot be supported by a wall, it should be adequately supported by other effective means.

3.4.20.   Putlogs which have one end supported by a wall should have at that end a plane supporting surface at least 10 cm (4 in) deep.

3.4.21.   The dimensions of the putlogs should be appropriate to the load to be borne by them.

3.4.22.   The distance between two consecutive putlogs on which a platform rests should be fixed with due regard to the anticipated load and the nature of the platform flooring.

3.4.23.   As a general rule the distance between two consecutive putlogs on which a platform rests should not exceed 1 m (3 ft 3 in) with planks less than 4 cm ($1^5/_8$ in) thick, 1.5 m (5 ft) with planks less than 5 cm (2 in) thick, and 2 m (6 ft 6 in) with planks at least 5 cm (2 in) thick.

3.4.24.   If putlogs are removed from a scaffold before it is dismantled, they should be replaced by a sufficient number of adequate braces.

Bracing

3.4.25.   Pole scaffolds that are tied to the building should be braced diagonally from top to bottom over the whole length.

3.4.26.   The bracing should be securely anchored to every ledger and upright at the crossing points.

3.4.27.   Independent pole scaffolds should be braced in the same way as tied scaffolds, and also crosswise.

3.4.28.   Bracing should be left in position until the last practicable moment in order to retain stability of the remaining scaffold.

Catch platforms

3.4.29.   If a pole scaffold is used as a catch platform for roof work, it should be properly anchored to the building.

3.5. Suspended scaffolds with manually operated platform

Outriggers

3.5.1.   Outriggers should be:

(a) of adequate strength and cross-section to ensure the solidity and stability of the scaffold;

(b) installed at right angles to the building face; and

(c) carefully spaced to suit the hangers or deck irons.

3.5.2.   The overhang of the outriggers from the building should be such that the platform is fixed to hang not more than 30 cm (12 in) from the building face, subject to the exception mentioned in paragraph 3.2.22.

3.5.3.   The outriggers should be securely anchored to the building by bolts or other equivalent means.

3.5.4.   Anchor bolts should be properly tightened and securely tie down the outrigger to the framework of the buildings.

3.5.5.   When outriggers are anchored by bags of ballast or other loose counterweights, the bags or counterweights should be securely lashed to the outriggers.

3.5.6.   Supporting hooks or roof irons should be forged from suitable steel, or equivalent material.

3.5.7.   Stop bolts should be placed at the end of each outrigger, and at the end of each supporting joist.

Suspension ropes

3.5.8.   Suspension ropes should:

(a) be made of high-grade manila fibre or other natural or synthetic fibre of equal quality, or steel wire; and

(b) have a safety factor of at least 10 for fibre and at least 6 for steel.

3.5.9. The upper ends of suspension ropes should terminate in a spliced loop or other equally satisfactory connection provided with a steel ring or eye, and the bolt should pass through the outrigger shackle and the ring and be secured by a nut.

3.5.10. Suspension ropes should pass through suitable pulley blocks so as to enable the platform to be raised and lowered safely.

3.5.11. Suspension ropes should be adequately protected against chafing.

3.5.12. Pulley blocks should be fastened to the platforms by the hangers.

Platforms

3.5.13. Platforms of suspended scaffolds with manually operated platform should not exceed 8 m (26 ft) in length or 60 cm (24 in) in width.

3.5.14. Platforms should:

(a) be suspended from two or more ropes or chains spaced not more than 3.5 m (12 ft) apart;

(b) be supported on ledgers in one piece resting on metal hangers attached to the suspension ropes or chains; and

(c) have an overhang not exceeding 75 cm (2 ft 6 in) beyond the hangers.

3.5.15. No intermediate rope should at any time be tauter than either of the end ropes.

3.5.16. Platforms should be supported on stirrups or hangers of mild steel or wrought iron of adequate cross-section.

3.5.17. Platform hangers should pass under the platform planks and be secured to them.

Operation

3.5.18. Not more than two workers should be employed on a suspended scaffold with manually operated platform at any one time.

3.5.19. Two or more suspended scaffolds with manually operated platform should not be combined by connecting them with planking or by other means.

3.5.20. When not in use a suspended scaffold with manually operated platform should be lashed to the building, or lowered to the ground and cleared of tools, other objects and rubbish.

3.5.21. Suspended scaffolds with manually operated platform should be tested before being used, by raising them a short distance under twice the safe working load.

3.5.22. Suspended scaffolds with manually operated platform on which the workers sit to work should be controlled or provided with devices to keep the platform at a distance of at least 45 cm (1 ft 6 in) from the wall, to prevent the workers from knocking their knees against the wall if the scaffold swings.

3.5.23. When a suspended scaffold with manually operated platform is not being raised or lowered, the hauling part of the fall rope should be made fast with a self-locking hitch to the lower block.

3.6. Suspended scaffolds with machine operated platform

Outriggers

3.6.1. Outriggers should comply with the requirements of paragraphs 3.5.1 to 3.5.4, 3.5.6 and 3.5.7.

3.6.2. Only under extremely exceptional circumstances should a counterweight be used as a means of securing the outriggers.

Suspension ropes

3.6.3. Only steel wire ropes [1] complying with the requirements of paragraphs 3.5.8 *(b)* and 3.5.9 should be used for suspended scaffolds with machine operated platform.

---

[1] Also referred to as "cables".

3.6.4. Suspension ropes should be of such length that at the lowest position of the platform there are at least two turns of rope on each drum.

3.6.5. Suspension ropes should be fastened to the outriggers vertically above the drum centres of the winches on the movable platforms.

3.6.6. The lower ends of suspension ropes should be securely fastened to the hoisting machines by clips, babbitting or other effective means.

Scaffolding (hoisting) machines

3.6.7. Scaffolding or hoisting machines should be so constructed and installed that their moving parts are readily accessible for inspection.

3.6.8. The frame of the hoisting machine should be securely fastened to the platform bearers by bolting or other effective means.

3.6.9. Winches on suspended scaffolds should be:

(a) of the self-braking type; or

(b) provided with a ratchet and pawl and a positive locking device such that the platform can be securely held at any level, and the pawl automatically engages when released from the hand control. When the pawl must be disengaged before lowering, an adequate safety device should be provided to prevent the reversing of the winch when the pawl is being disengaged.

3.6.10. If the power for raising and lowering is supplied from a motor, the motor should stop and automatically lock and securely hold the platform when the manual pressure on the starting switch or lever is released.

3.6.11. The moving parts of every scaffold machine should be inspected at least once every week.

3.6.12. When a scaffolding machine is moved from one place to another it should be inspected and overhauled before being used again.

Platforms

3.6.13. Platforms should comply with the requirements of paragraph 3.5.14.

3.6.14. Platforms should not exceed 8 m (26 ft) in length and 1.5 m (5 ft) in width.

Operation

3.6.15. While persons are working on a heavy suspended scaffold the suspension ropes should be secured by locking the winches or by other effective means.

3.6.16. Suspended scaffolds with machine operated platform should be prevented from swinging or knocking against the building by means of ties, spacing bars, etc.

3.6.17. When not in use suspended scaffolds with machine operated platform should:

(a) be cleared of all tools and other movable objects; and
(b) be securely lashed in position or lowered to the ground.

3.7. Outrigger, cantilever or jib scaffolds

3.7.1. Outrigger, cantilever or jib scaffolds should:

(a) be securely fixed and anchored from the inside;
(b) have outriggers of adequate length and cross-section to ensure their solidity and stability; and
(c) be properly braced and supported.

3.7.2. Only sound parts of a building which are sufficiently stable and sufficiently strong should be used as supports for scaffold parts.

3.7.3. If working platforms rest on bearers let into the wall, the bearers should be properly braced, go right through the wall and be securely fastened on the far side.

3.7.4. Outriggers should be spaced at intervals not exceeding 1.8 m (6 ft).

3.7.5. Platforms of outrigger scaffolds should be securely suspended from the outriggers by adequate hangers and supporting beams.

3.7.6. Platforms of outrigger scaffolds should not exceed 1.5 m (5 ft) in width.

3.7.7. Outrigger scaffolds should not be used for the storage of materials.

3.8. Ladder scaffolds

3.8.1. Ladder scaffolds should be used only:

(a) for light work requiring little material; and
(b) if they are suitable for the purpose.

Ladders

3.8.2. The ladders serving as the uprights of ladder scaffolds should either:

(a) be let into the ground to the necessary depth according to the nature of the soil; or
(b) be placed on sole-plates or boards so that the two uprights of each ladder rest evenly on the base, and be suitably fastened at the feet to prevent them from slipping.

3.8.3. If a ladder is used to extend another:

(a) the two ladders should overlap at least 1.5 m (5 ft);
(b) the upper ladder should be secured by two ladder hooks on a steel tie rod; and
(c) the join should be braced with straps.

Bracing

3.8.4. Single ladder scaffolds should be provided over their whole length with diagonal bracing extending over the whole height of every second bay.

3.8.5. The bracing of single ladder scaffolds should be securely fastened to the scaffold at every crossing point.

3.8.6. Double ladder scaffolds should be provided with sufficient diagonal bracing.

3.8.7. The bracing fastening of double ladder scaffolds should comply with the requirements of paragraph 3.8.5.

Anchorages

3.8.8. If ladder scaffolds are tied to the building:

(a) each ladder should be anchored to the building at each storey;

(b) the vertical distance between consecutive anchorages should not exceed 4.5 m (15 ft); and

(c) ladders should not extend more than 3 m (10 ft) above the highest anchorage.

3.9. Ladder jack scaffolds

3.9.1. Ladder jacks should only be used on ladders of sufficient strength to withstand safely the stress that a jack would impose.

3.9.2. Ladder jacks should not be used:

(a) at heights exceeding 6.5 m (22 ft); or

(b) on extension ladders.

3.9.3. Ladder jacks should be securely fastened to the ladder and should bear on the side rails.

3.9.4. Ladders used for ladder jack scaffolds should be prevented from being displaced by safety shoes, metal spikes or other effective means.

3.9.5. Not more than one person should work on a ladder jack scaffold at a time.

3.10. Bracket scaffolds

3.10.1. No bracket scaffold should be used unless the brackets are of suitable strength, are made of suitable metal, and are securely anchored in the wall by adequate bolts with washers and nuts.

3.10.2. Bracket scaffolds should only be used by workers such as carpenters, electricians and painters who do not require heavy equipment or materials.

3.10.3. The working platform of a bracket scaffold should not exceed 75 cm (2 ft 6 in) in width.

3.10.4. Brackets should be designed to withstand safely a load of at least 175 kg (400 lb) at the extreme outer end.

3.10.5. Brackets should be securely assembled by gussets and bolts.

3.10.6. Brackets should be provided at intervals not exceeding 3.5 m (12 ft).

3.11. Trestle scaffolds

3.11.1. No trestle scaffold should:

(a) be of more than two tiers; or
(b) exceed a height of 3 m (10 ft) from the ground or floor, or from the platform of a fixed scaffold; or
(c) be erected on a suspended scaffold.

3.11.2. Trestle scaffolds should be set up on firm and level bases.

3.11.3. The width of a trestle scaffold erected on a platform should be such as to leave sufficient unobstructed space on the platform for the transport of materials or the passage of persons.

3.11.4. Trestles should be firmly fixed so as to prevent displacement.

3.11.5. Trestles used for trestle scaffolds should be adequately braced to ensure rigidity and resist lateral thrusts.

3.11.6. Extension pieces should not be nailed on the legs of trestles to increase height.

3.11.7. Trestles should be so spaced as to ensure the stability of the scaffold.

3.11.8.   On extensible trestles the extension should be prevented from coming out of its seating.

3.12. Square scaffolds

3.12.1.   Squares should be securely assembled and braced to ensure adequate strength and rigidity.

3.12.2.   Squares should not have sides longer than 1.5 m (5 ft).

3.12.3.   Uprights of squares should not be longer than the sides.

3.12.4.   Square scaffolds should be set up on a firm and level base.

3.12.5.   On square scaffolds the squares should be set up at intervals not exceeding 1.5 m (5 ft).

3.12.6.   Square scaffolds should not be more than three tiers of squares high.

3.12.7.   The tiers should be securely fastened together by bracing.

3.12.8.   The top two tiers should rest on planking laid on the tier beneath.

3.13. Window jack scaffolds

3.13.1.   Window jack scaffolds should only be used for working at the window opening through which the jack is placed.

3.13.2.   Only one person should be on a window jack scaffold at any one time.

3.13.3.   Window jacks should be designed to withstand a working load of at least 90 kg (200 lb) with a safety factor of 4.

3.13.4.   Window jacks at adjacent windows should not be joined by planks.

3.13.5. Window jacks should not be used to support other kinds of scaffold.

3.14. Stages for dumping railway trucks

3.14.1. Discharge points on stages for dumping railway trucks should be protected by movable barriers, guard-rails and toe-boards and other effective means.

3.14.2. Dumping stages that are regularly used as passageways should be protected by railing complying with the requirements of paragraphs 2.6.1 to 2.6.5.

3.14.3. Alongside the rail tracks there should be a safe walkway for the workers which should be at least 60 cm (2 ft) wide.

3.14.4. Dumping stages should be strong enough to safely accommodate a set of loaded trucks extending over the whole length of the stage.

3.14.5. If a dumping stage cannot safely carry a locomotive or more than a certain number of trucks, adequate precautions should be taken to prevent the stage from being overloaded.

3.14.6. Rail tracks on dumping stages should comply with the relevant requirements of paragraphs 9.1.1 to 9.1.15.

3.15. Tubular metal scaffolds

General provisions

3.15.1. Tubular metal scaffolds should:

(a) be made of adequate material such as galvanised steel tubing; and

(b) be of adequate strength to hold the anticipated load with a safety factor of 4.

3.15.2. All vertical and horizontal members of tubular metal scaffolds should be securely fastened together.

3.15.3. Where necessary to prevent danger, tubular metal scaffolds should be adequately diagonally braced at suitable intervals in the lengthwise and crosswise directions.

3.15.4. Scaffolds with metal members should not be erected in dangerous proximity to any overhead electricity transmission line or electricity transmission equipment and should in all cases be at a distance of more than 5 m (16 ft 6 in) or such distance laid down by a competent authority.

3.15.5. Tubes or pipes for tubular scaffolds should be straight and free from rust, corrosion, indentations, and other defects.

3.15.6. The ends of tubes should be squared to ensure even bearing over the whole area of the section at joints and other connections.

3.15.7. Tubing should be of adequate size and strength for the load it will have to carry, and in no case less than 5 cm (2 in) (or other closely similar standard size) in outside diameter.

Uprights

3.15.8. Uprights should always be maintained in a vertical position.

3.15.9. Joints in uprights of tubular scaffolds should be:

(a) close to ledgers or other members capable of constraining them against lateral displacement; and

(b) staggered so that adjacent joints are not at the same level.

3.15.10. With the material commonly used, the distance between uprights should not exceed:

(a) 1.8 m (6 ft) for heavy-duty scaffolds capable of supporting 350 kg/m^2 (75 lb/ft^2);

(b) 2.3 m (7 ft 6 in) for light-duty scaffolds capable of supporting 125 kg/m^2 (25 lb/ft^2).

Ledgers

3.15.11. Ledgers should:

(a) extend over at least two spaces between uprights; and

(b) be securely fastened to each upright.

3.15.12.   Joints between ledgers should:

(a) be close to uprights; and
(b) not be directly one over another at different storeys.

3.15.13.   The vertical distance between ledgers should not exceed 2 m (6 ft 6 in).

3.15.14.   If a tubular metal scaffold is tied to the masonry construction, ledgers should be firmly secured to the masonry by clincher plates or other effective means.

3.15.15.   If working platforms are removed, all ledgers should be left in place to preserve the rigidity of the scaffold.

Putlogs

3.15.16.   On tubular metal scaffolds one putlog should be placed at each upright.

3.15.17.   The span of any putlog on tubular metal scaffolds should not exceed 1.5 m (5 ft).

3.15.18.   The distance between putlogs of heavy-duty tubular metal scaffolds should not exceed 90 cm (3 ft), and between putlogs of light-duty scaffolds, 1.15 m (3 ft 9 in).

3.15.19.   Where putlogs rest on a wall of a building they should have a plane supporting surface at least 10 cm (4 in) deep.

Anchorages

3.15.20.   Tubular metal scaffolds with a single row of uprights should be securely anchored to the wall of the building.

3.15.21.   The anchorage pattern should be such that:

(a) the anchorage tubes are fastened to the scaffold face at the junctions of uprights and ledgers;
(b) the anchorage tubes are securely fastened in the masonry;
(c) the first, the last and every second upright are anchored to the building through anchorage tubes fastened in a staggered pattern to every second ledger on a given upright.

Railings

3.15.22. Toe-boards should be bolted or properly fixed to uprights or adequately secured to platform planks.

3.15.23. Guard-rails should be secured to uprights by suitable connectors.

3.15.24. Diagonal bracing should not be used as guard-rails unless it provides protection equivalent to that provided under section 2.6.

Fastenings

3.15.25. Fastenings for connecting members of tubular metal scaffolds should:

(a) be made of drop-forged steel or equivalent material; and

(b) accurately embrace, over the whole area of their bearing surfaces, the members on which they are used.

3.15.26. Fastenings should not:

(a) cause deformation in the tube; or

(b) themselves undergo deformation.

3.15.27. Where the efficacy of fastenings depends on frictional grip, they should not be used to transmit primary tensile forces.

3.15.28. Fittings having screw threads and nuts should not be used unless each nut is fully engaged on the corresponding thread.

3.16. Mobile scaffolds

3.16.1. Scaffolds supported on wheels should be adequately braced and stiffened to prevent dangerous distortion in use.

3.16.2. Mobile scaffolds should only be used on a firm, level surface.

3.16.3. The height of mobile scaffolds should not exceed four times the lesser base dimension.

3.16.4. Ladders giving access to mobile scaffolds should be secured to the scaffold.

3.16.5. When a mobile scaffold is in use the castors or wheels should be adequately blocked.

3.16.6. No person should ride on a scaffold that is being moved.

3.16.7. All material and equipment which may fall should be removed from the platform before moving the mobile scaffold.

3.17. Boatswain's chairs, skips, etc.

3.17.1. A skip, basket, boatswain's chair or similar equipment should only be used as a suspended scaffold in exceptional circumstances for work of short duration, and under the supervision of a responsible person.

3.17.2. When a skip, basket, boatswain's chair or similar equipment is used as a scaffold:

(a) it should be supported by ropes having a safety factor of at least 10, based on the total load including the self weight of the skip, basket, or boatswain's chair;

(b) precautions should be taken to prevent the workers from falling out;

(c) workers should wear a safety belt secured to an individually fastened lifeline complying with the requirements of paragraphs 36.1.16 to 36.1.31.

3.17.3. When a skip is used as a suspended scaffold it should have a guard-rail installed at 1 m (3 ft 3 in) high.

3.17.4. When a basket is used as a suspended scaffold it should be at least 1 m (3 ft 3 in) deep.

3.17.5. When a skip or a basket is used as a suspended scaffold it should be carried by two strong iron bands which should be properly secured to it, continued round the sides and bottom, and have eyes in the iron to receive the ropes.

3.17.6. The total live load on a boatswain's chair raised and lowered by hand should not exceed 110 kg (250 lb).

3.17.7. Boatswain's chairs that are used by persons in a sitting position should have a back rail or rope and a front rail, rope or post not less than 25 cm (10 in) above the seat.

3.17.8. Boatswain's chairs that are used by persons in a standing position should be provided with a guard-rail and toe-board on all sides.

3.17.9. The seat of a boatswain's chair should:

(a) possess adequate strength and be firmly secured; and

(b) measure at least 45 × 25 cm (18 in × 10 in).

3.17.10. The suspension ropes of boatswain's chairs should be made of high-grade manila fibre or steel wire or equivalent material.

3.17.11. The suspension rope should be securely fastened to a firm overhead structure or passed through a pulley block similarly fastened and firmly secured to an easily accessible firm structure.

3.17.12. Fibre rope should not be used on boatswain's chairs when workers in them are using a blow torch or any open flame.

3.17.13. Boatswain's chairs should only be used provided a safety belt is worn and so fastened that the workers will be safely held if they fall out of the chair.

3.17.14. Before a boatswain's chair is taken into use the overhead supports and the tackle should be inspected.

3.18. Aerial-basket trucks [1]

Definition

3.18.1. By "aerial-basket" is meant a basket, bucket or skip raised on an articulated or telescopic boom that is mounted on a motor truck.

---

[1] Also called "boom platforms".

Trucks

3.18.2. Aerial-basket trucks should be constructed in conformity with the relevant requirements of section 10.2.

3.18.3. When necessary it should be possible to block parked trucks securely by means such as feet, outriggers, and jacks.

3.18.4. Locking devices should be provided to secure the boom and turntable during travel.

3.18.5. The turntable should lock automatically when the basket is in use.

3.18.6. The design should be such that in the case of motor or pump failure, the basket will remain in position.

3.18.7. The design should be such that if there is a failure in the hydraulic lines, descent will be limited to a safe speed.

3.18.8. Automatic cut-outs should be provided:

(a) to prevent the basket moving beyond set limits; and

(b) to prevent overloading if the basket fouls a fixed object.

Booms

3.18.9. Booms should have a safety factor of at least 25 per cent above the maximum rated load.

3.18.10. The aerial-basket boom should not be used as a derrick unless it has been specifically designed for this purpose; if so, it should be equipped with an appropriate safe load indicator.

Baskets

3.18.11. Aerial-baskets should:

(a) be at least 1 m (3 ft 3 in) deep and of such a design as to prevent danger to the user; and

(b) be electrically insulated, if necessary to prevent danger.

3.18.12. Aerial-baskets should be equipped with:

(a) a safety belt or safety harness for each occupant; and

(b) means for attaching the safety belt or safety harness to the basket or to the boom.

3.18.13. Aerial-baskets should bear a clear indication of the maximum rated load.

3.18.14. Safe means such as steps and guard-rails and toe-boards should be provided on the truck for climbing in and out of the bucket.

3.18.15. Basket controls should be protected against impact by position or by guards.

Inspection and maintenance of equipment

3.18.16. Aerial-basket trucks should be accompanied by the manufacturer's instructions for inspection and maintenance.

3.18.17. Aerial-basket trucks should be inspected and maintained in accordance with the manufacturer's instructions and the relevant requirements of section 10.5.

3.18.18. Aerial-basket trucks, including the boom, basket and accessories, should be inspected daily when in use.

3.18.19. In the course of each daily inspection the boom should be operated through one cycle.

3.18.20. Whenever necessary to prevent danger, trucks should be thoroughly overhauled.

3.18.21. Overhauls should include all parts of the hydraulic and lubricating systems.

Operation of equipment

3.18.22. Aerial-basket trucks should only be operated by competent persons.

3.18.23. Aerial-basket trucks should be operated in conformity with the relevant requirements of section 10.6.

3.18.24. An aerial-basket truck should not be moved with the boom raised before the operator has ensured that there will be no danger from obstructions or electrical conductors.

3.18.25. As far as practicable an aerial-basket truck should not be moved while any person is in the basket.

3.18.26.   As far as practicable trucks should only be used on firm, level and non-slippery ground.

3.18.27.   If a truck is used on an incline the axles should be kept as horizontal as possible so that the truck does not lean sideways.

3.18.28.   At work sites on public highways, trucks should be protected by means such as warning signs, lights and barricades.

3.18.29.   If a truck has to be used in darkness the working area should be adequately lighted.

3.18.30.   When an aerial-basket is used in the vicinity of aerial conductors the relevant requirements of section 17.8 should be complied with.

3.18.31.   When the aerial-basket is used for work on electrical installations, the work should only be done by competent electricians.

3.18.32.   Persons in aerial-baskets should not stand on the rim, or on boards placed across the rim, or use ladders.

3.18.33.   Persons in aerial-baskets should prevent tools and materials from falling out and refrain from throwing them out.

4. Ladders and stairs

4.1. General provisions

Construction

4.1.1. No wooden ladder having any rung which depends for its support solely on nails, spikes, screws or other similar fixing should be used.

4.1.2. Wooden ladders should be constructed with:

(a) uprights of adequate strength made of wood free from visible defects and having the grain of the wood running lengthwise; and

(b) rungs made of wood free from visible defects and mortised or rabbeted into the uprights.

4.1.3. Uprights and rungs of metal ladders should have a cross-section adequate to prevent dangerous deflection.

4.1.4. The intervals between rungs should be:

(a) equal; and
(b) not be less than 25 cm (10 in) or more than 35 cm (14 in).

4.1.5. Rungs of metal ladders should be corrugated or treated to prevent slipping.

4.1.6. Rungs of metal ladders should be kept clean so as to prevent them from becoming slippery.

4.1.7. If necessary to prevent danger, ladders should be provided with non-slip shoes, spikes or other devices to prevent slipping.

4.1.8. Wooden ladders should be provided with a sufficient number of steel cross-ties to ensure rigidity.

4.1.9. Portable ladders should not exceed 9 m (30 ft) in length.

4.1.10. Every ladder or run of ladders rising to a height exceeding 9 m (30 ft) should be provided with an intermediate landing or landings such that the interval between landings does not exceed 9 m (30 ft).

4.1.11. Landings should be:

(a) of suitable size; and

(b) protected by railings complying with the requirements of paragraphs 2.6.1 to 2.6.5.

Inspection, maintenance

4.1.12. Defective ladders that cannot be satisfactorily repaired should be destroyed.

4.1.13. Portable wooden ladders should be stored in a dry, well-ventilated place.

4.1.14. Wooden ladders should not be painted, but oiled or covered with clear varnish or transparent preservatives.

4.1.15. Metal ladders should be protected against corrosion by being coated with rust-proof paint or by other adequate means unless they are made of non-corrosive metals.

Use

4.1.16. Every ladder used as a means of communication should:

(a) rise at least 1 m (3 ft 3 in) above the highest point to be reached by any person using the ladder; or

(b) have one of the uprights continued to that height to serve as a handrail at the top.

4.1.17. Ladders should not stand on loose bricks or other loose packing but should have a level and firm footing so that they are equally supported on each upright.

4.1.18. Every ladder:

(a) should be securely fixed so that it cannot move from its top and bottom points of rest; or

(b) if it cannot be secured at the top, should be securely fastened at the base; or

(c) if fastening at the base is also impracticable, should have a man stationed at the foot to prevent slipping.

4.1.19.   Undue sagging of ladders should be prevented.

4.1.20.   Where a run of two or more ladders connects different floors:

(a) the ladders should be staggered; and

(b) a protective landing with the smallest practicable opening should be provided at each floor.

4.1.21.   A ladder having only one upright or a missing or dangerously defective rung should not be used.

4.1.22.   Roofers' and painters' ladders should not be used by workers in other trades.

4.1.23.   When it is placed in position the distance between the foot of a ladder and the base of the structure against which it rests should be about one-quarter of its length.

4.1.24.   Workers using ladders should:

(a) leave both hands free for climbing up and down;

(b) face the ladder;

(c) avoid wearing slippery boots or shoes; and

(d) avoid carrying heavy or bulky loads.

4.1.25.   If objects have to be carried on ladders, suitable means should be used for the purpose.

4.1.26.   A ladder should not be placed in front of a door that opens towards it unless the door is fastened open or is locked or guarded.

4.1.27.   A ladder should not be placed against a window frame unless the ladder is fitted with a board at the top so that the applied load is safely distributed over the frame.

4.1.28.   Metal ladders should not be used in the vicinity of exposed live electrical equipment.

4.1.29.   Adequate means should be provided to prevent the displacement of a ladder set up in a public thoroughfare or in any other place where persons, vehicles, etc. may accidentally collide with it.

4.1.30.   Where necessary to prevent danger, separate ladders should be used for ascent and descent.

4.2. Portable step ladders

4.2.1.   Portable step ladders should not exceed 6 m (20 ft) in length.

4.2.2.   Back legs of step ladders should be adequately braced.

4.2.3.   Step ladders exceeding 1.5 m (5 ft) in length should be equipped with two or more cross-ties.

4.2.4.   The spread between the front and back legs should be restrained by means of hinged metal flat bars or high-grade fibre rope or other effective means.

4.2.5.   When in the open position, treads of step ladders should be horizontal.

4.3. Portable trestle ladders

4.3.1.   Trestle ladders should not exceed 5.5 m (18 ft) in height.

4.3.2.   The spread between the front and back legs should be restrained by means of hinged metal flat bars or high-grade fibre rope or other effective means.

4.3.3.   The front and back legs should be joined at the top by bolted steel hinges of adequate dimensions or other effective means.

4.3.4.   Both legs of trestle ladders should be equipped with a sufficient number of steel cross-ties.

4.4. Extension ladders

4.4.1.   Extension ladders should not exceed 15 m (50 ft) in length.

4.4.2.   Extension ladders should be equipped with an effective lock and guide brackets by which the ladder can be extended, retracted or locked in any position.

4.4.3. The rungs of overlapping sections should coincide so as to form double treads.

4.4.4. Extension ladders should be equipped with one or more extension ropes.

4.4.5. Extension ropes should be securely anchored and run over suitable pulleys.

4.4.6. Extension ladders should have not more than two sliding extensions in addition to the base unit.

4.5. Mechanical ladders

4.5.1. By mechanical ladders is meant mechanically extensible ladders mounted on a wheeled frame.

4.5.2. Mechanical ladders should be equipped with:

(a) a working platform with guard-rails and toe-boards complying with the requirements of paragraphs 2.6.1 to 2.6.5; or

(b) a cage of heavy-gauge steel mesh.

4.5.3. If a mechanical ladder has no railed platform or cage, workers using it should be secured by a safety belt complying with the relevant requirements of Chapter 36.

4.5.4. Mechanical ladders should not be moved while a person is on them unless they have been specially designed to ensure that perfect stability is maintained during movement.

4.6. Fixed ladders

4.6.1. Fixed ladders installed out of doors should be made of steel.

4.6.2. Fixed steel ladders should comply with the requirements of section 4.1 concerning metal ladders.

4.6.3. Uprights of fixed ladders should be at least 40 cm (16 in) apart.

4.6.4. Where practicable, fixed ladders should be set at an angle of 15° to the vertical.

4.6.5.   The clearance at the back of the rungs should be at least 15 cm (6 in).

4.6.6.   There should be no obstruction within 75 cm (30 in) of the face of the ladder.

4.6.7.   At the sides there should be at least 7.5 cm (3 in) clearance between the ladder and the nearest fixed object.

4.6.8.   When it is necessary for a ladder to pass closely through a hole in a platform or a floor, the edges of the hole should be padded so as to prevent injury to persons using the ladder.

4.6.9.   Runs of fixed ladders should not exceed 9 m (30 ft) in length.

4.6.10.   Landing platforms should be provided for each 9 m (30 ft) or fraction thereof.

4.6.11.   As far as possible, runs should be staggered.

4.6.12.   Runs of fixed ladders from which a person could fall from more than 6 m (20 ft) should be enclosed in a cage of heavy-gauge steel mesh or hoops.

4.6.13.   Fixed ladders should be firmly bolted or welded in position.

4.7. Stairs

4.7.1.   Stairs should be of adequate strength to withstand safely the loads that they will have to carry.

4.7.2.   Stairs used for the purpose of construction work should have a clear width of at least 60 cm (2 ft).

4.7.3.   Stairs made of perforated material should not have openings exceeding 1.2 cm (½ in) in width.

4.7.4.   No step of a stairway should depend for its support solely on nails, spikes, screws or other similar fixing.

4.7.5.   No stairway with missing or dangerously defective steps should be used.

4.7.6. Stairs with more than five steps should:

(a) be provided on any open side with guard-rails and toe-boards complying with the requirements of paragraphs 2.6.1 to 2.6.5;

(b) if more than 1.2 m (4 ft) wide, be provided on both sides with an adequate handrail, or if this is not practicable, with an adequate hand rope;

(c) if not more than 1.2 m (4 ft) wide, be provided on one side with an adequate handrail, or if this is not practicable, with an adequate hand rope.

4.7.7. Every stairway that is at an angle of less than 30° from the vertical should be provided with a secure handhold at the top landing place, either by extending one upright for at least 1 m (3 ft 3 in) or by other effective means.

4.7.8. Movable and removable stairs should be adequately secured in the position of use.

4.7.9. No flight of stairs used for the purpose of operations being carried on should have an unbroken vertical rise of more than 3.5 m (12 ft).

4.7.10. In all buildings permanent stairs should be constructed as soon as practicable and should comply with the relevant provisions of this Code.

4.7.11. When work on a building has progressed to a height of more than 18 m (60 ft) above the ground and it has not been practicable to construct the permanent stairs, a sufficient number of stairs should be provided to ensure safe access to working levels.

5. Lifting appliances

5.1. General provisions

Maximum safe working load

5.1.1.   Adequate steps should be taken to ascertain the maximum safe working load or loads of every lifting appliance.

5.1.2.   The maximum safe working load should be marked:

(a) upon every winch and pulley block used in the hoisting or lowering of any load;

(b) upon every derrick pole or mast; and

(c) upon every crane.

5.1.3.   In the case of a crane fitted with a derricking jib, the safe working load at various radii of the jib should be marked upon it, or otherwise clearly indicated.

5.1.4.   The maximum safe working load or loads should be incised or stamped in a legible and durable manner, or marked at a conspicuous place in some other legible and durable manner.

Installation

5.1.5.   Fixed lifting appliances should be installed:

(a) by competent persons;

(b) so that they cannot be displaced by the load, vibration or other influences;

(c) so that the operator is not exposed to danger from loads, ropes or drums; and

(d) so that the operator can either see over the zone of operations or communicate with all loading and unloading points by telephone, signals or other adequate means.

5.1.6.   Adequate clearance should be provided between moving parts and loads of lifting appliances and:

(a) fixed objects such as walls and posts; and

(b) electrical conductors.

5.1.7.   Every lifting appliance should be adequately supported.

5.1.8.   When lifting appliances are exposed to wind loadings, they should be given sufficient additional strength, stability and rigidity to withstand such loadings safely.

5.1.9.   No structural alterations or repairs should be made on any part of a lifting appliance that affects the safety of the appliance without the permission of the competent person.

Operator's cab or cabin

5.1.10.   The operator of every lifting appliance in outdoor service should be provided with a cab or cabin which should:

(a) be made of fire-resisting material;

(b) have a suitable seat and a foot rest including protection from vibration;

(c) afford the operator an adequate view of the area of operation;

(d) afford the necessary access to working parts in the cab;

(e) afford the operator adequate protection against the weather;

(f) be adequately ventilated;

(g) be adequately heated in cold weather;

(h) be free from the risk of the operator coming into contact with the heating appliance when operating the controls;

(i) be provided with a suitable fire extinguisher.

Controls

5.1.11.   Controls of lifting appliances should be:

(a) so situated that the operator at his stand or seat has ample room for operation and an unrestricted view, and remains clear of the load and ropes, and that no load passes over him;

(b) provided, where necessary, with a suitable locking device to prevent accidental movement or displacement.

5.1.12.   Control handles should, as far as possible, move in the direction of the resultant load movement, or clockwise for lifting and anti-clockwise for lowering.

5.1.13.   The stroke of hand levers should not exceed 60 cm (2 ft).

5.1.14.   The stroke of pedals should not exceed 15 cm (6 in).

5.1.15. Pedals should have a non-slip surface.

5.1.16. Lifting appliances should be equipped with devices that:

(a) prevent the load from over-running;

(b) prevent the load from moving if the power fails.

Brakes

5.1.17. Brakes should at all times be capable of performing the task for which they were originally designed.

5.1.18. If necessary to prevent danger, brakes should be provided with a locking device.

5.1.19. Brakes should act without shock or delay.

5.1.20. Brakes should be provided with simple and easily accessible means of adjustment.

5.1.21. Brakes operated by hand should not require a force greater than 16 kg (35 lb) at the handle.

5.1.22. Brakes operated by foot should not require a force greater than 32 kg (70 lb) on the pedal.

Buckets

5.1.23. Tip-up buckets should be equipped with a device that effectively prevents accidental tipping.

Winches and drums

5.1.24. Winches and drums used on lifting appliances should comply with the requirements of section 5.11.

Ropes and tackle

5.1.25. Ropes and tackle used on lifting appliances should comply with the relevant requirements of Chapter 6.

Inspection, maintenance

5.1.26. Lifting appliances should be thoroughly inspected and, if necessary to prevent danger, tested at least once a year by a competent person.

5.1.27. Every part of the structure, working gear and anchoring and fixing appliances of every crane, hoist and winch, and of all other lifting appliances and tackle, should, as far as the construction permits, be inspected in position at least once in every week by the operator or other competent person.

5.1.28. Means of communication such as telephone and signalling equipment should be tested before each working period.

Operation

5.1.29. Every crane driver or lifting appliance operator should be a competent person.

5.1.30. No person under 18 years of age should be in control of any lifting appliance, including any scaffold winch, or give signals to the operator.

5.1.31. Precautions should be taken to prevent lifting appliances from being set in motion by unauthorised persons.

5.1.32. The operation of lifting appliances should be governed by signals, in conformity with the requirements of section 1.8.

5.1.33. The lifting appliance operator's attention should not be distracted while he is working.

5.1.34. No crane, hoist, winch or other lifting appliance, or any part of such appliance, should, except for testing purposes, be loaded beyond the safe working load.

5.1.35. During hoisting operations effective precautions should be taken to prevent any person from standing or passing under the load.

5.1.36. Operators should not leave lifting appliances unattended with power on or with a load suspended.

5.1.37. No person should ride on a suspended load or on any lifting appliance not authorised by the competent authority for the conveyance of persons.

5.1.38. Every part of a load in course of being hoisted or lowered should be adequately suspended and supported so as to prevent danger.

5.1.39.   Every receptacle used for hoisting bricks, tiles, slates or other material should be so enclosed as to prevent the fall of any of the material.

5.1.40.   If loose materials or loaded wheelbarrows are placed directly on a platform for raising or lowering, the platform should be enclosed.

5.1.41.   Materials should not be raised, lowered or slewed in such a way as to cause sudden jerks.

5.1.42.   In hoisting a barrow, the wheel should not be used as a means of support unless efficient steps are taken to prevent the axle from slipping out of the bearings.

5.1.43.   If necessary to prevent danger, long objects such as planks and girders should be guided with a tag line, while being raised or lowered.

5.1.44.   At landings, workers should not be obliged to lean out into empty space for loading and unloading.

5.1.45.   The hoisting of loads at points where there is a regular flow of traffic should be carried out in an enclosed space, or if this is impracticable (e.g. in the case of bulky objects), measures should be taken to hold up or divert the traffic for the time being.

5.1.46.   Adequate steps should be taken to prevent a load in course of being hoisted or lowered from coming into contact with any objects in such a manner that part of the load or object may become displaced.

5.1.47.   Appliances should be provided and used for guiding when raising or lowering heavy loads (for example, prefabricated parts) to avoid crushing of the hands in these operations.

5.2. Hoists

Shafts and towers

5.2.1.   Hoist shafts should be provided with rigid panels or other adequate fencing:

(a) at the ground level on all sides; and
(b) at all other levels on all sides to which access is provided.

5.2.2.   The walls of hoist shafts, except at approaches, should extend at least 2 m (6 ft 6 in) above the floor, platform or other place to which access is provided.

5.2.3.   Approaches to hoists should be provided with substantial gates or the like which:

(a) should be gridded for visibility;
(b) should be at least 2 m (6 ft 6 in) high; and
(c) should be equipped with a device which requires the gate to be closed before the platform leaves the landing and prevents the gate from being opened unless the hoist platform is at the landing.

5.2.4.   Approaches to hoists should be adequately lit.

5.2.5.   The guides of hoist platforms should offer sufficient resistance to bending and, in the case of jamming by a safety catch, to buckling.

5.2.6.   Overhead sheave beams and their supports should be capable of holding the combined maximum live and dead loads that they will have to carry with a safety factor of at least 5.

5.2.7. A clearance should be provided above the highest stopping place high enough to allow sufficient unobstructed travel of the cage or platform in case of overwinding.   A clear space should also be provided below the lower stopping place.

5.2.8.   Adequate covering should be provided above the top of hoist shafts to prevent material falling down them.

5.2.9.   Outdoor hoist towers should be erected on adequately firm foundations, and securely braced, guyed and anchored.

5.2.10.   A ladderway complying with the relevant requirements of Chapter 4 should extend from the bottom to the top of outdoor hoist towers, if no other ladderway exists within easy reach.

Engines

5.2.11. Hoisting engines should be of ample capacity to control the heaviest load that they will have to move.

5.2.12. All gearing on hoisting engines should be securely enclosed.

5.2.13. If necessary to prevent danger, steam piping of hoisting engines should be adequately protected against accidental contact.

5.2.14. Electrical equipment of hoisting engines should be effectively earthed.

5.2.15. Hoists should be provided with devices that stop the hoisting engine as soon as the platform reaches its highest stopping place.

5.2.16. Hoisting engines should be protected by a substantial cover against the weather and falling objects.

5.2.17. If hoisting engines are set up in a public thoroughfare, they should be completely enclosed.

5.2.18. Exhaust steam pipes should discharge so that the steam cannot scald anyone or obstruct the operator's view.

5.2.19. It should not be possible to reverse the motion of the hoist without first bringing it to rest.

5.2.20. It should not be possible to set in motion from the platform a hoist not designed for the conveyance of persons.

5.2.21. Pawls and ratchet wheels with which the pawl must be disengaged before the platform is lowered should not be used.

Wire ropes [1]

5.2.22. Steel wire ropes should be used for the suspension of the platform and should comply with the requirements of section 6.2.

5.2.23. Hoist wire ropes should have a safety factor of at least six times the maximum load.

---

[1] Also referred to as "cables".

5.2.24. If two or more wire ropes are used, the load should be equally distributed between them.

5.2.25. Each suspension wire rope should be in one piece.

5.2.26. The rope ends should be fastened to the platform attachment by splicing and tight binding with steel wire, by sealing or by clamping with the aid of rope clamps; wherever possible, properly applied thimbles should be used.

5.2.27. Drum anchorages of suspension wire ropes should be adequate and secure.

5.2.28. Wire ropes should be long enough to leave at least two turns on the drum when the cage or platform is at its lowest position.

5.2.29. The diameter of the pulleys or drums should not be less than 20 times the diameter of the wire rope used.

Platforms

5.2.30. Hoist platforms should be capable of supporting the maximum load that they will have to carry with a safety factor of at least 3.

5.2.31. Hoist platforms should be equipped with safety gear that will hold the platform with the maximum load if the hoisting rope breaks.

5.2.32. On platforms, wheelbarrows and trucks should be efficiently blocked in a safe position.

5.2.33. If workers have to enter the cage or go on the platform at landings there should be a locking arrangement preventing the cage or platform from moving while any worker is in or on it.

5.2.34. On sides not used for loading and unloading, hoist platforms should be provided with toe-boards and enclosures of wire mesh or other suitable material to prevent the fall of parts of loads.

5.2.35. Where necessary to prevent danger from falling objects, hoist platforms should be provided with adequate covering.

Counterweights

5.2.36.   Counterweights consisting of an assemblage of several parts should be made of specially constructed parts rigidly connected together.

5.2.37.   Counterweights should run in guides.

Landings

5.2.38.   Adequate platforms complying with the requirements of section 3.2 should be provided at all levels used by workers.

Notices

5.2.39.   The following notices should be posted up conspicuously and in very legible characters:

(a) on all hoists:

   (i) *on the platform:* the carrying capacity in kilograms or other appropriate standard unit of weight; and

   (ii) *on the hoisting engine:* the lifting capacity in kilograms or other appropriate standard unit of weight;

(b) on hoists authorised or certified for the conveyance of persons:
      on the platform or cage: the maximum number of persons to be carried at one time;

(c) on hoists for goods only:
      on every approach to the hoist: prohibition of use by persons.

Inspection, maintenance

5.2.40.   No hoist should be used unless it has been examined and tested by a competent person and a certificate of such test and examination has been issued by that person.

5.2.41.   Examinations and tests referred to in paragraph 5.2.40 should be repeated:

(a) at regular intervals which should be prescribed by the competent authority; and

(b) after every substantial alteration or repair and every re-erection.

Operation

5.2.42.　When the platform is at rest the brake should be applied automatically.

5.2.43.　During loading and unloading the platform should be blocked by catches or other devices in addition to the brake.

5.2.44.　Pipes and other long objects should be securely lashed to prevent them from catching in the tower as the platform moves.

Concrete bucket hoists

5.2.45.　Concrete bucket hoists should comply with the requirements of section 25.4.

Conveyance of persons

5.2.46.　No hoist should be used for the conveyance of persons unless:

(a) such use has been authorised by the competent authority; or
(b) the hoist complies with the conditions laid down for the installation and operation of lifts used for the conveyance of persons in industrial undertakings.

5.3. Cranes

Structure

5.3.1.　Stress-bearing structural members of cranes that are also subject to shock should be constructed of mild steel or other equally suitable material.

5.3.2.　Cranes should be so designed and constructed that all parts can be safety lubricated, inspected and repaired.

5.3.3.　Access to and egress from the operator's stand should be safe in any position of the crane.

Erection of cranes

5.3.4.　The erection of cranes should be supervised by a competent person.

Anchorage, ballast

5.3.5.   Every fixed crane should either be securely anchored or be adequately weighted by suitable ballast firmly secured to ensure stability.

5.3.6.   When a crane is weighted by ballast a diagram showing the position and size of the counterweights should be posted up in the operator's cab.

5.3.7.   Loose material such as bricks and stones should not be used as ballast for cranes.

Jib cranes

5.3.8.   On jib cranes:

(a) means such as a stirrup piece should be provided to prevent the rope from coming off the pulley at the end of the jib; and

(b) when the jib is fixed and cannot be lowered to the ground, access to the pulley should be provided by a ladder protected by guard-rails and toe-boards complying with the requirements of paragraphs 2.6.1 to 2.6.5.

5.3.9.   When the jib of a jib crane is at the maximum radius there should not be less than two dead turns of rope on the derricking drum.

Slewing cranes [1]

5.3.10.   Slewing cranes with power-operated slewing mechanism should be equipped with a power-operated brake for the slewing motion.

5.3.11.   Slewing cranes that slew on wheels should be equipped with means that prevent them from overturning if a wheel breaks.

5.3.12.   Precautions should be taken to prevent workers from being crushed between rotating parts of cranes and the carriage.

Scotch derrick cranes

5.3.13.   The jib of a Scotch derrick crane should not be erected between the back stays of the crane.

---

[1] Or pivoting cranes.

Power control on electric cranes

5.3.14. It should only be possible to lower a load with the motor switched on, and so that the rated number of revolutions of the motor is not exceeded.

5.3.15. Electric cranes should be equipped with at least one switch by which power for all movements can be cut off on all poles from the operator's stand.

5.3.16. Main control switches should be protected against unauthorised use.

5.3.17. The hoisting mechanism should automatically stop if even only one phase of the power fails.

5.3.18. Electric cranes should be equipped with overload protection devices that act on:

(a) the hoisting mechanism;
(b) the jib-raising and lowering mechanism; and
(c) the jib crab[1] if there is one.

5.3.19. After the overload prevention device has acted, it should be possible to lower the load and pull in the crab.

5.3.20. It should be possible to render the jib-locking device inoperative by a switch so that the jib can be pulled in.

5.3.21. Limit switches should limit:

(a) the upward travel of the hook;
(b) the upward and downward travel of the jib; and
(c) the backward and forward travel of the crab.

5.3.22. After limit switches have acted, travel in the opposite direction should be possible.

5.3.23. After the limit switch for the hook has acted, it should not be possible to lower the jib.

5.3.24. It should only be possible to switch on the power when all controls of electric cranes are at zero.

5.3.25. Controls and switches should not be blocked.

---

[1] Mobile hoisting drum.

5.3.26. Before leaving the crane the operator should switch off the power.

Load and radius indicators

5.3.27. Power-driven jib cranes should be fitted with an automatic indicator which:

(*a*) indicates clearly to the operator when the load being moved approaches the maximum safe working load of the crane at any inclination of the jib; and

(*b*) gives an efficient signal when the load being moved is in excess of the maximum safe working load of the crane at any inclination of the jib.

5.3.28. No crane with either a fixed jib or a derricking jib should be used unless a plate has been fixed in the driver's cabin which indicates to the driver the safe working loads that correspond to the following conditions when the crane is standing on a firm horizontal base: the length of the jib; the radius of operation; operation with or without stabilising screw-jacks.

5.3.29. The crane driver must take into account the angles of inclination shown on the jib indicator in the case of derricking jib cranes.

5.3.30. The maximum permitted angle of inclination of the crane jib must be clearly marked on the crane.

5.3.31. When the change-speed gear is used for the hoisting machine, the safe working load corresponding to each speed should be prominently indicated and visible to the operator.

Inspection, testing

5.3.32. No crane should be used unless a competent person has:

(*a*) inspected and tested it; and

(*b*) furnished a certificate specifying:

    (i) the maximum safe working load at the various radii at which the jib can be worked; and

(ii) in the case of a crane with a derricking jib, the maximum radius at which the jib may be worked.

5.3.33. The examinations and tests referred to in paragraph 5.3.32 should be repeated:

(a) at regular intervals which should be prescribed by the competent authority;
(b) after all substantial alterations or repairs to the crane; and
(c) after every re-erection.

5.3.34. The safe working load at any radius specified in the most recent certificate should not be more than the load which the crane has stood at that radius during the application of the test.

5.3.35. Before being put into use for the first time, jib cranes with a variable radius should undergo tests:

(a) of stability; and
(b) of all movements such as travel, slewing, raising and lowering the load, braking the crane and braking the load.

5.3.36. Cranes should be tested for anchorage by the imposition on each anchorage of the maximum uplift or pull exerted either:

(a) by a load of 25 per cent above the maximum load to be lifted by the crane as erected; or
(b) by a lesser load arranged to exert an equivalent pull on the anchorage.

5.3.37. If the pull applied by the test to any anchorage is less than 25 per cent in excess of the pull which would be exerted by the maximum safe working load, a loading diagram appropriate to the crane anchorage should be affixed in a position where it can readily be seen by the crane operator.

5.3.38. If a crane has been exposed to weather likely to have affected its stability, the anchorage and ballast should be examined and the crane found to be safe before it is used again.

Operation

5.3.39. Cranes should not be used to pull out fixed objects, lift with a slanting pull, drag objects or move vehicles.

5.3.40. No load which lies in the angle between the back stays of a Scotch derrick crane should be moved by that crane.

5.3.41. No crane should be used in weather likely to endanger its stability.

5.3.42. When a load is thought to approach the maximum safe working load, the operator should proceed with extreme caution and pay special attention to the warning and signalling devices.

5.3.43. Jib cranes should not be moved or operated in dangerous proximity to electric power lines.

5.3.44. While cranes are working no person should be in the area of operations other than those engaged in the operations.

5.3.45. Where more than one crane is required to lift or lower one load:

(a) arrangements should be such that none of the cranes will at any time be loaded beyond its safe working load or be rendered unstable in the hoisting or lowering of the load; and
(b) a person should be specially appointed to co-ordinate the operation of the cranes working together.

5.4. Travelling cranes

Tracks

5.4.1. Tracks of travelling cranes should be of adequate section, properly laid, firm, level, of adequate bearing capacity, and have an even running surface.

5.4.2. All rails on which a travelling crane moves should, unless other adequate steps are taken to ensure the proper junction of the rails and to prevent any material alteration in their gauge:

(a) be jointed by fish-plates or double chairs; and
(b) be securely fastened to sleepers.

5.4.3. The ends of tracks should be provided with buffers or stop blocks of adequate shock-absorbing capacity.

Passageways

5.4.4. On every stage, gantry or other place on which a crane moves there should in so far as practicable be maintained at every position of the crane an unobstructed passageway of a width of at least 60 cm (2 ft) between the moving parts of the crane and the fixed parts or edge of such stage, gantry or place.

5.4.5. If at any time it is impracticable to maintain a passageway of a width of at least 60 cm (2 ft) at any place or point, precautions should be taken to prevent the access of any person to such place or point at such time.

Clearances

5.4.6. Between track-mounted cranes and other objects along the crane track there should always be a sufficient clearance to prevent danger.

Electric crane tracks

5.4.7. Track rails of electric cranes should:

(a) be protected against excessive contact voltages; and
(b) be electrically bonded and earthed.

5.4.8. Crane contact rails should be so installed or protected that accidental contact is prevented in normal operation.

5.4.9. Channels for contact rails should be drained.

Trolley lines

5.4.10. Overhead trolley lines for electric travelling cranes should comply with the relevant requirements of section 17.6.

5.4.11. Where necessary to prevent danger, trolley lines should be protected against contact with stacks of material, moving equipment, falling objects, etc.

Crane travelling structures

5.4.12. Track-mounted cranes should be provided with effective brakes for the travelling motion.

5.4.13. Track-mounted cranes should be provided with:

(a) struts that prevent their collapse in the case of wheel breakage and are so positioned that they can act as foot guards;

(b) anchoring devices such as rail clamps that will prevent them from being overturned by wind pressure; and

(c) a device for removing obstacles, snow and ice from the rails.

5.4.14. It should be possible to cut off the power from the contact rail on all poles by means of an easily accessible switch.

5.4.15. If two or more cranes are supplied from the same contact rail, each should have a disconnecting switch directly behind the current collector.

5.5. Overhead travelling cranes

Tracks

5.5.1. Tracks of overhead travelling cranes should comply with the requirements of paragraphs 5.4.1 to 5.4.11.

5.5.2. Tracks of overhead travelling cranes should not be used as walkways.

5.5.3. Alongside the track of overhead travelling cranes there should be a walkway complying with the requirements of section 3.3.

5.5.4. If a safe walkway, referred to in paragraph 5.5.3, cannot be provided, recesses should be provided at suitable intervals.

5.5.5. It should be possible to cut off the power from the crane on all poles by means of a suitably located remote switch.

5.5.6. The switch referred to in paragraph 5.5.5 should:

(a) be of a type allowing it to be locked in the open position;

(b) be combined with lamps or other devices showing whether the power is on or off.

5.5.7. In the cab there should be a switch by which the power can be cut off from the track.

5.5.8. Where necessary to prevent danger, adequate precautions should be taken to prevent contact between contact rails and:

(a) the operator as he goes to and from his cab; and

(b) the hoisting rope or hook.

Crane construction

5.5.9. Overhead travelling cranes should comply with the requirements of paragraphs 5.4.12 to 5.4.15.

5.5.10. Overhead travelling cranes should be provided with automatic switches limiting the travel of:

(a) the crab on the bridge girder;

(b) the hook upwards and downwards; and

(c) the crane on the track.

5.5.11. Automatic limit switches of overhead travelling cranes should not prevent movement in the opposite direction.

5.5.12. On overhead travelling cranes, mechanical and electrical equipment that is not directly accessible from a track walkway or a bridge walkway should be accessible from a suitable working platform.

5.5.13. Overhead travelling crane drives and crab drives should have power-operated brakes.

5.5.14. Overhead travelling cranes should be provided with a device that warns the operator when the wind is approaching the maximum safe velocity.

5.5.15. Adequate precautions should be taken to prevent the fall of gears, wheels and other parts.

5.5.16. There should be safe means of access to operators' cabs such as gangways, ramps, stairs and fixed ladders complying, in particular, with the requirements of sections 3.3 and 4.6.

5.5.17. Operators of overhead travelling cranes should be protected against:

(a) harmful fumes, gases and other atmospheric contaminants; and

(b) harmful radiations.

5.5.18. Overhead travelling cranes should be provided with shock-absorbing bumpers such as hydraulic buffers.

Bridges

5.5.19. At least one of the bridge girders should be provided with a walkway complying with the requirements of section 3.3.

5.5.20. There should be convenient and safe means of access to and egress from bridge walkways.

5.5.21. Openings for access and egress should be so placed that users will not be endangered by the crab.

Operation

5.5.22. In high winds overhead travelling cranes should be securely anchored.

5.6. Tower slewing cranes

General provisions

5.6.1. On tower slewing cranes, the turntable should be so constructed that objects cannot become caught in the gear.

5.6.2. Tower slewing cranes on rails should comply with the relevant requirements of sections 5.4 and 5.5.

5.6.3. Tower slewing cranes with power-driven slewing mechanism should have a brake for the slewing motion.

5.6.4. If a crab operates on the jib of a tower slewing crane, the crab drive should be so constructed that a braked crab cannot slip even on an icy or greasy track.

5.6.5. Tower slewing cranes should be equipped with loud warning devices.

5.6.6. Counterweight jibs that are loaded with ballast after erection should be provided with a walkway complying with the requirements of section 3.3.

5.6.7. Means of access to elevated operators' cabs should comply with the requirements of paragraph 5.5.16.

5.6.8. Trailing cables should run over a drum that automatically winds and unwinds them.

Erection

5.6.9. When tower cranes are being erected:

(a) the area of operations should be fenced off or guarded; and
(b) the erection crew should wear safety belts complying with the relevant requirements of Chapter 36.

5.6.10. Tower cranes should not be erected in high winds, snow-storms or icy conditions.

Ballast, counterweights

5.6.11. Operating instructions should indicate the weight and position of the ballast.

5.6.12. The ballast or counterweight should be firmly secured in position.

5.6.13. If the ballast and counterweight vary with the height of the tower and the radius of the jib, a table should be provided on the crane indicating the ballast and counterweight required for different tower heights and jib radii.

Operation

5.6.14. Tower slewing cranes should be operated in accordance with the manufacturer's instructions.

5.6.15. The manufacturer's operating instructions should be kept on the crane.

5.6.16. A grab should not be used as a hook.

5.6.17. At the close of work, the manufacturer's instructions for the safety of the crane against wind pressure should be observed.

5.6.18. Tower slewing cranes should not be operated in high winds or violent storms and, where necessary, should be provided with anemometers.

5.7. Monorail hoists or single track-mounted hoists [1]

General provisions

5.7.1.   Monorail hoists should have power-operated brakes.

5.7.2.   Monorail hoists should be so installed that they cannot fall if the king-bolt of the suspension gear breaks.

5.7.3.   Monorail hoists exposed to high winds should be provided with protection against the wind, such as rail anchors.

5.7.4.   Monorail hoists should be equipped with a loud warning device.

5.7.5.   Loose parts such as brake weights and signal bells should be secured against falling.

5.7.6.   When monorail hoists are designed to allow workers to travel on the platform, this platform should be protected by guard-rails and toe-boards complying with the requirements of paragraphs 2.6.1 to 2.6.5.

5.7.7.   The ends of the crab or mobile platform should be provided with securely anchored strong buffers.

5.7.8.   Contact rails should be so laid or protected that accidental contact with them is not possible.

5.7.9.   The protected conductor should be laid in the same way as the contact rail.

5.7.10.   The tracks of monorail hoists should be protected against excessive contact voltages.

Power control

5.7.11.   The upward travel of the lifting device should be limited by a limit switch.

5.7.12.   Limit switches should allow travel in the opposite direction.

5.7.13.   It should be possible to cut off main contact rails

---

[1] Also called "underhung trolley" or "underhung crab".

and mobile conductors on all poles by means of a mains switch in an easily accessible position.

5.7.14.   On monorail hoists there should be a hoist switch by which the electrical equipment of the crab or mobile platform can be cut off on all poles from the operator's stand.

5.7.15.   If a number of monorail hoists are supplied from the same contact rail, there should be isolating switches in an easily accessible position immediately behind the current collectors, unless the hoist switch itself is immediately behind the current collectors.

5.7.16.   If monorail hoists are operated from the ground:

(a) control chains, operating cords, etc. should be prevented from becoming entangled;

(b) control appliances should be so constructed that they automatically cut off the power when they are released.

Operator's cab

5.7.17.   The operator's cab of a monorail hoist should comply with the requirements of paragraph 5.1.10.

5.7.18.   At least one means of access to the operator's cab should be a stairway with a platform.

5.7.19.   Means of access should be so installed that the users are not endangered by the moving crab.

5.7.20.   It should be possible to leave the crab safely in any of its positions.

5.8. Derricks

Stiff-leg derricks

5.8.1.   Derricks should be erected on a firm base and adequately secured against displacement.

5.8.2.   Derrick ropes, sheaves and bridles should be so arranged that they do not foul parts of the derrick or any obstructions during slewing or otherwise.

5.8.3. Suitable devices should be used to prevent masts from lifting out of their seatings.

5.8.4. Electric derricks should be effectively earthed from the sole-plate or framework.

5.8.5. Counterweights should be so arranged that they do not subject the backstays, sleepers or pivots to excessive strain.

5.8.6. When derricks are mounted on wheels:

(a) a rigid member should be used to maintain the correct distance between the feet; and

(b) they should be equipped with struts to prevent them from dropping if a wheel breaks or the derrick is derailed.

Guy derricks

5.8.7. The mast of guy derricks should be supported by six top guys spaced approximately equally.

5.8.8. Where the guys of a guy derrick crane cannot be fixed at approximately equal spacing, such other measures should be taken as will ensure the safety of the crane.

5.8.9. Guy ropes of derricks should be equipped with a stretching screw or turnbuckle or other device to regulate the tension.

Operation of derricks

5.8.10. Derricks, guys and loads should not be allowed to come into dangerous proximity to electrical conductors.

5.8.11. Gudgeon pins, sheave pins and foot bearings should be lubricated frequently.

5.8.12. When a derrick is not in use, the boom should be lowered to prevent it from swinging.

5.8.13. Truck-mounted derricks should have their maximum loads clearly indicated to avoid danger of overturning the truck.

5.9. A-frames, sheer legs

5.9.1. A-frames should be erected on a firm, level base.

5.9.2. A-frames should be adequately guyed and anchored to prevent overturning or displacement.

5.9.3. A-frames should be of adequate strength to support safely the maximum loads that they will have to carry.

5.9.4. Legs of A-frames should be of steel or other suitable metal or sound straight-grained timber or at least equivalent material.

5.9.5. The legs should be prevented from spreading.

5.9.6. The angle between the legs and the horizontal should be at least 75°.

5.9.7. The legs should be securely fastened together with bolts and locked nuts.

5.9.8. The top hoisting block should be suspended by steel wire rope.

5.10. Gin poles, gin wheels

5.10.1. Gin poles should:

(a) be straight;
(b) consist of steel or other suitable metal or straight-grained timber free from knots;
(c) be adequately guyed and anchored;
(d) be vertical or raked slightly towards the load; and
(e) be of adequate strength for the loads that they will be required to move.

5.10.2. Gin poles should not be spliced.

5.10.3. Gin poles should be adequately fastened at the foot to prevent displacement in operation.

5.10.4. When a gin pole is used on a scaffold, it should be secured by ropes in such a way that the load cannot knock against the scaffold.

5.10.5. Gin poles that are moved from place to place and re-erected should not be taken into use again before the pole, lifting ropes, guys, blocks and other parts have been inspected, and the whole appliance has been tested under load.

5.10.6. When platforms or skips are hoisted by gin poles, adequate precautions should be taken to prevent them from spinning and to provide for proper landing.

5.10.7. Pulley blocks or gin wheels supported by a beam or gallows should be firmly secured to the beam or gallows.

5.10.8. The supporting beam or gallows should be of adequate strength for the purpose for which it is being used.

5.10.9. The supporting beam or gallows should be adequately secured against displacement by lashing, counterweights or other effective means.

5.11. Winches

General provisions

5.11.1. All parts of the framework of winches should be of metal.

5.11.2. Frames of winches should be securely anchored to substantial foundations.

5.11.3. Where necessary to protect operators against the weather, breakage of ropes or flying objects, winches should be roofed over or otherwise adequately screened.

5.11.4. The means of protection of winches should not obstruct the operator's view.

5.11.5. Winches should be equipped with an acoustic signalling device.

5.11.6. On every winch the control lever should be provided with a suitable locking device.

Drums

5.11.7. Winch drums should:

(a) have a smooth winding surface;

(b) have a diameter at least 20 times the diameter of the rope wound on them; and

(c) have flanges extending at least two rope diameters beyond the last turn of rope.

5.11.8. There should be a fastening point for the rope at each end of a winch drum or barrel, and the rope should be securely fastened to the drum or barrel.

5.11.9. The attachment of the rope to the winch drum should be capable of holding at least three times the maximum working load.

5.11.10. When winch drums are grooved:

(a) the radius of the grooves should be approximately the same as, but not less than, the radius of the rope; and

(b) the pitch of the grooves should not be less than the diameter of the rope.

Hand-operated winches

5.11.11. Hand-operated winches should be so constructed that the maximum effort to be applied by any one person at the handle or handles when the winch is lifting its maximum safe working load should not exceed:

(a) as a general rule, 10 kg (22 lb);

(b) in any case, 16 kg (36 lb).

5.11.12. Hand-operated winches should be provided with ratchet wheels on the drum shafts and locking pawls, or self-locking worm gears, to prevent reversing while loads are being hoisted.

5.11.13. Hand-operated winches should be provided with effective braking devices for controlling the lowering of the loads.

5.11.14. Crank handles for hand-operated winches should:

(a) be so constructed that they do not turn while the loads are being lowered by means of the brake; or

(b) be removed before the loads are lowered.

5.11.15. Detachable crank handles for hand-operated winches should be secured against accidental removal.

5.12. Jacks

Construction

5.12.1. Jacks should be of such construction that the load:
(a) will remain supported in any position;
(b) cannot be lowered inadvertently;
(c) will not slip off the bearing surface.

5.12.2. The rated capacity of every jack should be clearly incised or stamped on it.

5.12.3. Every jack should have a stop or other effective means of preventing overtravel.

5.12.4. Electric jacks should be provided with automatic limit switches at the top and bottom limits of travel.

5.12.5. Hydraulic and pneumatic jacks should be provided with devices that will prevent the load from falling suddenly if the cylinder containing the liquid or air is damaged.

5.12.6. Screw and rack jacks should be provided with devices that prevent the screw or the rack from coming out of its seating.

Use

5.12.7. Every jack should be accompanied by instructions for its safe use and proper maintenance.

5.12.8. When lifting objects with jacks, the jacks should be:
(a) set on solid footings;
(b) centred properly for the lifts; and
(c) so placed that they can be operated without obstruction.

5.12.9. Jacks should be tested under load at suitable intervals.

6. Ropes[1], chains and accessories

6.1. General provisions

6.1.1. All new or reconditioned chains, rings, hooks, shackles, swivels and pulley blocks used for hoisting or lowering or as a means of suspension should have been tested and marked in plain figures and letters with the safe working load before being put into use.

6.1.2. All chains, rings, hooks, shackles and swivels used in hoisting or lowering or as a means of suspension should be examined and tested before being put into use.

6.1.3. All cables or ropes used on hoisting appliances for raising or lowering materials should be long enough to leave at least two turns on the drum at every operating position of the appliance.

6.1.4. No rope should be used over a grooved drum or pulley if its diameter exceeds the pitch of the drum grooves or the width of the pulley groove.

6.1.5. Every hoisting or derricking rope or chain should be securely fastened to the drum of the crane, crab or winch with which it is used.

6.1.6. Sharp edges of a load should not be in contact with slings, ropes or chains.

6.1.7. All chains, ropes, slings and other gear used for hoisting or lowering or as a means of suspension should be periodically inspected by a competent person and this person's findings should be entered on a certificate or in a special register.

6.1.8. Chains and gear, such as rings, hooks, shackles and swivels on lifting appliances should, if appropriate, be given suitable heat treatment at regular intervals in accordance with rules which should be laid down by the competent authority.

---

[1] Including wire ropes or "cables".

6.1.9.   When not in use, ropes, chains and accessories should be stored under cover in clean, dry, well-ventilated places where they are protected against corrosion or other damage.

6.1.10.   As far as practicable ropes, chains and accessories in storage should be so arranged that items with the same maximum safe working load are grouped together.

6.1.11.   No gear used for attachment or as a means of suspension should be loaded beyond its maximum safe working load.

6.2. Wire ropes

6.2.1.   Wire ropes for lifting appliances should:

(a) be made of sound steel wire;

(b) have a safety factor related to the method of use but at least 3.5 times the maximum load;

(c) consist of one length; and

(d) be free from knots, kinks and frayed sections.

6.2.2.   In order to prevent kinking, twisting or untwisting of new wire rope, the rope should:

(a) when received in coils, be uncoiled by rolling the coils like hoops on level surfaces, and straightened out before being put on sheaves; or

(b) when received in reels, be unwound by:

   (i) rolling the reels along the ground;

   (ii) pulling the ends of reels mounted horizontally on spindles or vertically on turntables.

6.2.3.   Ends of wire rope should be seized or otherwise secured to prevent the strands from coming loose.

6.2.4.   Splices and fastenings of wire rope should be carefully examined at regular intervals, and clips or clamps tightened if they show signs of loosening.

6.2.5.   In order to keep wire ropes pliable and prevent rust, the ropes should, if practicable, be treated at regular intervals with suitable lubricants free from acids or alkalis.

6.2.6. Reverse bends in wire rope should be avoided as far as practicable.

6.2.7. Wire ropes should be regularly inspected and be replaced in the case of extensive wear, corrosion, breakage of wires or other dangerous defects.

6.2.8. Wire ropes should be fastened to hooks, tongs, etc. by secure means.

6.2.9. Eye splices and loops of wire ropes should be provided with thimbles.

6.2.10. If wire ropes other than ropes for lifting appliances and inclined transport are joined by splicing, the strength of the splice should at least be equal to that of the wire ropes.

6.2.11. Wire ropes should be cut with a suitable tool and using a soft hammer, not a hard hammer or an axe.

6.2.12. When a wire rope is used, the diameter of the pulleys or drums should not be less than 20 times the diameter of the wire rope.

6.3. Fibre ropes

6.3.1. Fibre ropes for lifting appliances should be of high-grade manila fibre or other natural or synthetic fibre of equal characteristics and quality.

6.3.2. Before being put into use and while in use, at intervals to be determined according to the nature of the work but not to exceed three months, fibre ropes for lifting appliances should be examined for abrasions, broken fibres, cuts, fraying, displacement of yarns or strands, variation in size or roundness of strands, internal wear between strands, deterioration of fibre, discoloration and other defects.

6.3.3. Splices of fibre ropes should not be repaired but the faulty splice should be cut out and replaced by a new splice.

6.3.4. Fibre ropes should not be exposed to abrasion from rough surfaces, grit, sand, etc., or to corrosion by acids, alkalis, fumes, etc., or to great heat.

6.3.5. Fibre ropes should be threaded only through blocks that:

(a) have no sharp or rough edges or undue projections; and
(b) have sheaves with grooves at least as wide as the diameter of the rope and free from roughness.

6.3.6. Wet fibre ropes should not be allowed to freeze.

6.3.7. Fibre ropes should not be lubricated.

6.3.8. In storage fibre ropes should:

(a) be hung on suitably shaped wooden pegs, galvanised hooks, or the like, separately from metal gear; and
(b) be protected against rodents.

6.4. Chains

6.4.1. Chains used for lifting and transport appliances should be withdrawn from use whenever:

(a) the chains have become unsafe from overloading or through faulty or improper heat treatment;
(b) any one link of the chain has stretched more than 5 per cent of its length; or
(c) other external defects are evident.

6.4.2. Chains should only be repaired by properly qualified persons having suitable equipment for the purpose.

6.4.3. Chains that are wound on drums or pass over sheaves should be lubricated at frequent and regular intervals.

6.4.4. Chains should not be:

(a) hammered to straighten links or force them into position;
(b) crossed, twisted, kinked or knotted;
(c) dragged from under loads;
(d) dropped from a height;
(e) used to roll loads over;
(f) subjected to shock loads.

6.4.5. Joining broken chains by wiring links together, by inserting bolts between links, or by passing one link through another and inserting a bolt or nail to hold it, should be prohibited.

6.4.6. Chains should be examined for stretch, wear, gouge marks, cracks and open welds at frequent intervals.

6.4.7. When individual links of lifting or transport chains show excessive wear or are bent, cut, gouged or cracked, they should be cut out and replaced.

6.5. Slings

6.5.1. All slings should be made of chains, wire ropes or fibre ropes of adequate strength to withstand the stresses to which they will be subjected.

6.5.2. Rings, hooks, swivels and end links of hoisting chains should be made of the same material as the chains.

6.5.3. Tables showing the maximum safe working loads for slings at various angles should be displayed in conspicuous places.

6.5.4. Workers using slings should be familiar with the tables referred to in paragraph 6.5.3.

6.5.5. Slings that show evidence of cuts, excessive wear, distortion or other dangerous defects should be withdrawn from use.

6.5.6. Wire rope slings should be kept well lubricated.

6.5.7. Where necessary to prevent sharp bends in slings, corners of loads should be adequately padded.

6.5.8. When multiple slings are used the load should be distributed equally among the ropes as far as practicable.

6.5.9. Where double or multiple slings are used for hoisting or lowering purposes, the upper ends of the slings should be connected by means of a shackle or ring and not be put separately into a lifting hook.

6.5.10. When bulky objects are being raised or lowered, the proper number of slings should be selected to ensure stability and also to support the weight of the load.

6.6. Pulley blocks

6.6.1. Pulley blocks should be made of metal resistant to shock (mild steel or equivalent material).

6.6.2. Axles of pulleys should be made of metal of suitable quality and of adequate dimensions.

6.6.3. The diameter of the pulley should be at least 20 times the diameter of the rope to be used.

6.6.4. The axle in the blocks should be capable of being lubricated. A suitable lubricating device should be provided for this purpose if practicable.

6.6.5. Regular and adequate lubrication of axles of pulleys should be ensured.

6.6.6. The sheaves and housing of blocks should be so constructed that the rope cannot become caught between the sheave and the sides of the block.

6.6.7. The grooves in the sheaves should be such that the rope cannot be damaged in the sheave.

6.6.8. Badly worn blocks should be taken out of use.

6.6.9. Blocks designed for use with fibre rope should not be used with wire rope.

6.6.10. Pulleys within reach of workers should be provided with a guard that effectively prevents a hand from being drawn in.

6.7. Hooks

6.7.1. Hooks for lifting appliances should be of forged steel or equivalent material.

6.7.2. Hooks should be provided with a safety catch or be so shaped as to prevent the load from accidentally slipping off.

6.7.3. If necessary to prevent danger, hooks should be provided with a hand rope (tag line) long enough to enable workers engaged in loading or unloading operations to keep clear.

6.7.4.   Parts of hooks liable to come into contact with ropes or chains during hoisting operations should have no sharp edges.

6.8. Shackles

6.8.1.   Shackles used for joining lines should have a breaking strength at least 1.5 times that of the lines joined.

6.8.2.   Shackles used for hanging blocks should have a breaking strength at least twice that of the pulling lines.

6.8.3.   Shackles used for hanging blocks should have the pins secured by locked nuts or other suitable means of equal safety.

6.8.4.   Shackle pins should be secured by keys or wire, unless bolts are employed.

7. Conveyors

7.1. General provisions

Construction, installation

7.1.1. Conveyors should be so constructed and installed as to avoid hazardous points between moving and stationary parts or objects.

7.1.2. If a walkway is provided along a belt conveyor, it should:

(a) be at least 60 cm (2 ft) wide;

(b) be kept clear of obstructions;

(c) be provided with guard-rails and toe-boards complying with the requirements of paragraphs 2.6.1 to 2.6.5 if it is more than 1.5 m (5 ft) above the ground.

7.1.3. When workers have to cross over conveyors, regular crossing facilities adequately lighted and affording safe passage should be provided if necessary to prevent danger.

7.1.4. When conveyors that are not entirely enclosed cross over places where workers are employed or pass beneath, sheet or screen guards should be provided to catch any material that might fall from the conveyors.

7.1.5. Power-driven conveyors should be provided at loading and unloading stations, at drive and take-up ends, and at other convenient places, if necessary to prevent danger, with devices for stopping the conveyor machinery in an emergency.

7.1.6. Adequate fencing should be provided at transfer points.

7.1.7. Where two or more conveyors are operated together, the controlling devices should be so arranged that no conveyor can feed on to a stopped conveyor.

7.1.8. Conveyors that carry loads up inclines should be provided with mechanical devices that will prevent the machinery from reversing and carrying the loads back towards the loading point in the event of the power being cut off.

7.1.9. Where the tops of hoppers for feeding conveyors are less than 90 cm (3 ft) above the floor, the openings should be adequately guarded.

7.1.10. Conveyors should be provided with automatic and continuous lubrication systems, or with lubricating facilities such that they can be safely oiled and greased.

7.1.11. Conveyors should be so designed that belts and drums can be cleaned safely.

7.1.12. Belt conveyors should be provided with guards at the nips of belts, rollers, and driving, reversing and tensioning sprockets.

7.1.13. Where necessary to prevent danger, a device should be provided at each end of the belt of a belt conveyor of more than 5 m (16 ft) in length to put the mechanism into neutral.

7.1.14. Screw conveyors should be enclosed at all times. The cover should not be removed until the conveyor is stopped.

Inspection, maintenance

7.1.15. Conveyors should be thoroughly inspected at suitable intervals.

7.1.16. Conveyors should not be repaired while in motion.

7.1.17. Rollers should not be lubricated while the belt is in motion unless lubrication can be carried out in conformity with accepted standards of safety.

Operation

7.1.18. Conveyor operations should be governed by signals in accordance with the requirements of section 1.8.

7.1.19. Workers should not ride on conveyors unless they are designed to carry persons.

7.1.20. Conveyors should not be overloaded so as to cause danger from falling material.

7.1.21. Employees should not be permitted to go under moving conveyors to clean up spilled material unless the mechanism is completely enclosed or fenced, including the return flight.

7.1.22. When a conveyor is discharging into a bunker or hopper, the feeding conveyor should be equipped with an overload switch.

7.2. Bucket elevators

7.2.1. On bucket elevators the nip points of ropes, chains and belts should be effectively guarded.

7.2.2. In traffic zones, bucket chains should be fenced or enclosed.

7.2.3. On bucket elevators with a lift of more than 5 m (16 ft), precautions should be taken to prevent the buckets from running back.

7.2.4. If necessary to prevent danger, precautions should be taken to prevent persons from being injured by falling material.

7.2.5. Loading points should be so arranged that persons cannot be caught by the moving buckets or in the machinery.

7.2.6. On enclosed bucket elevators, inspection openings should be kept closed and locked.

7.2.7. Bucket elevators should only be adjusted or repaired when they are stopped and precautions have been taken to prevent them from being set in motion.

7.2.8. No worker should climb the bucket ladder while the elevator is in motion.

7.2.9. No worker should remove heavy objects from a bucket while the elevator is in motion.

7.2.10. No worker should pass under the ladder while the elevator is in motion.

8. Aerial cableways [1]

8.1. General provisions

8.1.1.   All bearing parts of aerial cableways should be designed to withstand the maximum loads to which they will be subjected with a safety factor related to the method of run, but at least 3.5 times the maximum stresses, taking into account not only the load but also the secondary stresses due to wind, speed of operation, deflection, etc.

8.1.2.   Adequate side clearances should be provided between aerial cableways and fixed objects such as ramps, platforms and chutes.

8.1.3.   At places where persons work or pass, moving cables or ropes and the nip points on sheaves should be adequately enclosed or fenced.

8.1.4.   Tip-up buckets should be provided with locking devices that prevent inadvertent tipping.

8.1.5.   If workers have to pass from conveyances to supports, devices to enable this to be done safely should be provided on both.

8.1.6.   If work has to be done regularly on conveyances and supports, working platforms complying with the requirements of section 3.2 should be provided if necessary to prevent danger.

8.1.7.   If access to the tops of supports is regularly required, ladders equipped with cages should be provided and should comply with the relevant requirements of Chapter 4.

8.1.8.   At places where aerial cableways pass over passageways, workplaces, traffic routes, etc., adequate means should be provided to catch any falling objects or material.

---

[1] Also called "aerial ropeways".

8.1.9. The top platform should be provided with guard-rails and toe-boards complying with the requirements of paragraphs 2.6.1 to 2.6.5.

8.1.10. If necessary to prevent danger, special precautions should be taken to protect aerial cableways against forest fires.

Stations

8.1.11. Every station should be provided with effective means for communicating with all the others.

8.1.12. The maximum load to be carried by the conveyances should be posted up at every station.

8.1.13. Stations should be provided with adequate lighting.

8.1.14. In cold weather stations should be heated.

Engines

8.1.15. Engines should be equipped with:

(a) a speed governor that prevents the maximum safe speed from being exceeded;

(b) a device that stops them when the tensioning weights of the cables reach their limiting positions; and

(c) if necessary to prevent danger, a brake that will stop them if the supporting cable becomes slack.

8.1.16. Compressed-air operated brakes should be so constructed as to stop the cableway in the event of a compressed air failure.

8.1.17. No unauthorised person should be allowed in the engine room.

Towers

8.1.18. Supporting towers should be adequately guyed.

8.1.19. Supporting towers should be provided with safe means of access such as ladderways for repairs and inspection purposes.

8.1.20. Supporting towers should be high enough to give the moving equipment and loads a safe clearance over any workers employed below.

8.2. Aerial cableways carrying persons

General provisions

8.2.1. Aerial cableways carrying persons should comply with the requirements of section 8.1 and in addition with the requirements of this section.

8.2.2. Stations should be provided with searchlights by which the arrival and departure of conveyances can be watched.

8.2.3. Stations should be provided with a sufficient number of suitable fire extinguishers.

8.2.4. The supporting cables should have a safety factor of at least 3.5 and traction cables a safety factor of at least 4.

8.2.5. There should be means of rescuing passengers from conveyances in the event of a breakdown.

Engines

8.2.6. The building housing the engine should be of fire-resistant construction.

8.2.7. The winch should be driven by an electric motor.

8.2.8. The engine should be equipped with:

(a) a speed indicator; and
(b) an automatic overtravel prevention device.

8.2.9. It should be possible to run the winch at a slow speed for inspection purposes.

8.2.10. In addition to the electric motor there should be a thermal engine or a manual appliance by which conveyances can be brought to a station in an emergency.

8.2.11. Flat belts should not be used to transmit the power from a thermal engine.

8.2.12. No cast-iron gear wheels should be used.

8.2.13.   The traction installation should be provided with:

(a) a hand brake; and

(b) an automatic safety brake.

8.2.14.   The safety brake should act:

(a) if the engine stops;

(b) if the maximum safe speed is dangerously exceeded;

(c) when a conveyance reaches the travel limiting device;

(d) if the travel limiting device is inoperative.

8.2.15.   The operator's place should have a seat from which he can see the movements of the conveyances and the indicating and control instruments and operate the main switch to stop the cableway in an emergency.

Conveyances

8.2.16.   Conveyances for persons should be able to communicate with the engine room by telephone or other signalling equipment.

8.2.17.   In the event of a prolonged stoppage, provision should be made to enable passengers to leave a conveyance safely.

8.2.18.   Cage or cabin doors should not open outwards.

8.2.19.   Windows in cages or cabins should be of safety glass or equivalent material.

8.2.20.   Cages or cabins should be enclosed on all sides up to a height of at least 1.1 m (3 ft 6 in) and be provided with a roof.

8.2.21.   The cage or cabin should be so suspended from the carriage that it is horizontal when it stops.

8.2.22.   The load of the conveyances should be distributed equally over all the running wheels.

8.2.23.   The conveyances should be so attached to the traction rope that no wheel is pulled off the supporting rope.

8.2.24.   At all points on the cableway there should be an adequate clearance between the cage or cabin carrying the maximum load and the ground or usual snow level.

8.3. Inspection and maintenance of aerial cableways

8.3.1. Aerial cableways should be inspected and tested:

(a) before being taken into use for the first time;

(b) after any substantial alteration; and

(c) at suitable regular intervals.

8.3.2. Without prejudice to the requirements of paragraph 8.3.1, aerial cableways should be inspected:

(a) after any abnormal incident; and

(b) after any high wind, violent storm, or other natural occurrence likely to damage them.

8.3.3. Means of telephonic or radiotelephonic communication should be tested before a conveyance carrying persons leaves a station.

8.3.4. All working parts and safety devices such as brakes, overtravel preventers and speed governors should be tested before the first journey every day.

8.3.5. During inspection and repair operations:

(a) the engine driver should be in constant attendance at his post;

(b) the workers should, if necessary to prevent danger, wear a safety belt complying with the relevant requirements of Chapter 36.

8.3.6. Work on moving parts of the installation, on tensioning devices, in tensioning-weight pits and under tensioning weights should only be done when:

(a) the engine has been stopped; and

(b) precautions have been taken to prevent the engine from being inadvertently started up.

8.3.7. Pits and tensioning-weight towers should be kept clean, adequately ventilated and free from snow in cases where this may impair their operation.

8.3.8. All wire ropes and moving parts should be kept well lubricated.

8.4. Operation of aerial cableways

8.4.1.   Operators should be 21 years of age and should have been found fit by special medical examination.

8.4.2.   The operation of aerial cableways should be governed by signals in conformity with the requirements of section 1.8.

8.4.3.   Loads in conveyances should be adequately secured against displacement, falling over or falling out.

8.4.4.   No person should travel in a conveyance moving by gravity.

8.4.5.   No persons should be carried together with flammable or explosive substances.

8.4.6.   After a heavy snowfall or heavy frost or in a thick fog, a cableway should be operated without a load for one journey.

8.4.7.   A cableway should not be operated in high winds or violent storms.

8.4.8.   On aerial cableways not authorised for passenger transport, persons should only travel for inspection and repair purposes.

9. Railways

9.1. Railways with locomotive haulage

Track, general requirements

9.1.1. Construction railway tracks should be so laid and equipped with due regard to the gradients, curves, bearing capacity of the ground and the weight and speed of the traffic that they cannot cause danger.

9.1.2. All rails on which any locomotive, truck or wagon moves should:

(a) have an even running surface, be adequately supported and be of adequate section;

(b) be joined by fish-plates or double chairs, or other effective means;

(c) be securely fastened to sleepers or bearers, or be provided with other effective means of preventing any dangerous variations in the gauge;

(d) be supported on a surface sufficiently firm to prevent dangerous movement of the rails;

(e) be laid in straight lines or in curves of radii such that locomotives, trucks and wagons can move smoothly and safely; and

(f) be provided with adequate stops or buffers at every track end.

9.1.3. Sidings should be provided with derailers to protect the main track against runaways.

9.1.4. Turntables should be capable of being blocked.

9.1.5. Guard-rails should be provided at:

(a) curves on bridges and trestles; and
(b) switches.

9.1.6. As far as practicable, switch handles should move parallel to the tracks.

Clearances

9.1.7. There should be adequate overhead and side clearance between rail tracks and adjacent structures or objects.

9.1.8. At any place where the provision of adequate side clearance is not practicable, effective audible or visual signals should be provided to warn of the approach of locomotives or other rolling stock.

9.1.9. At any place where the provision of adequate overhead clearance is not practicable, effective audible or visual signals should be provided to warn locomotive drivers and other persons travelling on rolling stock.

9.1.10. When tracks run alongside excavations they should be at a safe distance from the edge, taking into account its solidity.

9.1.11. No material should be piled in dangerous proximity to tracks.

9.1.12. Sufficient clear space to allow safe working should be maintained around power-driven capstans or haulage winches.

9.1.13. Level footing clear of obstructions should be provided at places where:

(a) brakemen usually get on and off trains to throw switches, set brakes, or for other purposes; and

(b) trains are usually inspected.

Walkways

9.1.14. Every gantry, bridge, trestle, or other elevated structure carrying track should be provided with a safe walkway.

9.1.15. If necessary to prevent danger, walkways, referred to in paragraph 9.1.14, on the outside of the track should be protected by guard-rails and toe-boards complying with requirements of paragraphs 2.6.1 to 2.6.5.

Level crossings

9.1.16. Level crossings should:

(a) not be concealed by bushes, stacks of materials, etc.;

(b) be guarded or provided with adequate warning signs on both sides; and

(c) have the roadbed made flush with the rails by planking or other effective means.

Locomotives and other rolling stock

9.1.17.   Locomotives and self-propelled vehicles (speeders) should comply with national laws and regulations as regards materials, design, construction and equipment.

9.1.18.   Locomotives and self-propelled vehicles should be tested by a competent person in accordance with requirements which should be laid down by the competent authority before being taken into use.

9.1.19.   Rolling stock should be of suitable material and design, sound construction and adequate strength and be equipped with such devices as brakes, lights, whistles, sanders, guards, handholds, footholds and couplers as are necessary to prevent danger.

9.1.20.   Rolling stock used for transporting workers should:

(a) have sides at least 1 m (3 ft 3 in) high and have a top cover available for inclement weather;

(b) provide a fixed seat for each worker carried;

(c) have an easily accessible emergency brake;

(d) if necessary to prevent danger, have enclosed racks or boxes for tools; and

(e) if necessary for reasons of health or comfort, be heated and ventilated.

9.1.21.   Dump cars should be provided with a locking device for the container which prevents accidental tipping.

9.1.22.   Trucks that are moved alone should be provided with effective brakes unless they can be safely braked by hand.

9.1.23.   Trucks that are braked with sprags should have a device that prevents the sprag from slipping.

9.1.24.   Brake wagons should have a safe stand for the brakesman.

Inspection, maintenance

9.1.25. Railway tracks, installations, locomotives, self-propelled vehicles and rolling stock should be inspected at suitable intervals.

9.1.26. Locomotives and self-propelled vehicles should be inspected before the beginning of each shift.

9.1.27. Look-out men should be posted and all other necessary precautions taken to protect men repairing or otherwise working on tracks.

9.1.28. Rolling stock should not be inspected or lubricated while in motion.

General operating provisions

9.1.29. Construction railways should be operated in accordance with a code of rules approved by the competent authority.

9.1.30. The code should provide for adequate despatching and signalling systems.

9.1.31. Only competent persons should drive locomotives and self-propelled vehicles.

9.1.32. Trains controlled by hand brakes should carry a sufficient number of brakemen.

9.1.33. Persons carried as passengers on rolling stock should only ride on the fixed seats provided for them.

9.1.34. No person should ride on the buffers, running boards, or other unsafe parts of locomotives and other rolling stock.

9.1.35. Steam locomotives left unattended while under pressure should be securely blocked.

9.1.36. No worker should:

(a) remain unnecessarily on the tracks;
(b) crawl under vehicles;
(c) go between moving vehicles;
(d) pull vehicles by the front end or the buffers; or
(e) brake vehicles by pushing against them.

9.1.37. Adequate precautions should be taken to prevent collisions at level crossings and other places where it is possible for road traffic to move on the track.

9.1.38. Cars carrying passengers should not:

(a) normally be pushed by a locomotive; and
(b) be coupled with loaded trucks on steep gradients.

9.1.39. No passenger should get on or off a moving vehicle.

9.1.40. No flammable liquids should be carried on cars carrying passengers.

9.1.41. Explosives should be transported in conformity with the requirements of section 22.2.

Movement of vehicles

9.1.42. Adequate precautions should be taken to prevent stationary vehicles from moving inadvertently.

9.1.43. Trucks moved by hand should be pushed and not pulled.

9.1.44. Single trucks and sets of trucks should not travel unaccompanied.

9.1.45. No power-driven capstan or haulage winch should be set in motion for moving rolling stock until adequate warning has been given by effective audible or visual signals to all persons who might be endangered by the movements.

9.1.46. Only competent persons should operate power-driven capstans and haulage winches.

9.1.47. If vehicles are moved by animals:

(a) the animals should remain at the side of the track;
(b) it should be possible to detach the animals quickly on inclines and when otherwise necessary;
(c) the animals should be detached on inclines exceeding 1 in 100; and
(d) the driver should not be on the track or between the haulage rope and the track.

9.1.48. A sufficient number of suitable sprags or scotches should be provided for controlling the movement of trucks or wagons.

9.1.49. Suitable equipment should be provided for rerailing derailed rolling stock.

Loading and unloading vehicles

9.1.50. When a vehicle is tipping material over the edge or an excavation, embankment or earthwork, adequate precautions should be taken to prevent it from falling over the edge.

9.1.51. When dump cars are being loaded, it should be ensured that the locking device for the container is securely closed.

9.1.52. Long objects such as rails and masts should not be loaded on to dump cars.

9.1.53. No person should be in dangerous proximity to trucks while their doors are being opened, or they are being unloaded or dumped, or are automatically discharging.

9.1.54. When adhesive material such as wet sand and clay is being dumped, adequate precautions should be taken to prevent the truck from overturning.

9.1.55. No person should remain on any rolling stock while it is being loaded with loose material by means of a grab, excavator or similar appliance.

9.2. Inclined rope railways

General provisions

9.2.1. Every station should be equipped with means for communicating with all the others.

9.2.2. The maximum number of units to be coupled together should be posted at all stations.

9.2.3. The hauling appliance, such as a winch or drum, should have an effective brake.

9.2.4. When necessary to prevent danger, the nip point between in-running ropes and drums should be protected.

9.2.5. Haulage winches should comply with the requirements of section 5.11.

9.2.6. Haulage ropes on rope railways should have a safety factor of at least 3.5 for transport of material and 4 for transport of persons.

9.2.7. Rolling stock on rope railways should have effective brakes and a brakeman's stand.

9.2.8. Rolling stock should be provided with an effective mechanical drag to retard the descent in the case of mechanical failure.

9.2.9. Where workers have to be conveyed by inclined rope railways, the cars should:

(a) be provided with a device that will effectively hold them if the haulage rope breaks, or a coupling or a drawbar breaks or becomes detached;

(b) have level seats with handholds; and

(c) afford adequate protection against the weather and falling objects.

9.2.10. Haulage ropes on inclined rope railways conveying workers should be inspected at least once every day.

9.2.11. If necessary to prevent danger, workplaces and passageways under inclined rope railways should be protected against:

(a) falling goods or objects; and

(b) the consequences of breakage of the haulage rope.

Track, equipment

9.2.12. Rope railways should have stout barriers at loading points.

9.2.13. Rope railways should have:

(a) an adequate dead-end siding at the bottom for runaways;

(b) an effective derailing device; and

(c) if necessary to prevent danger, an effective automatic vehicle arrestor.

9.2.14.   Alongside rope railways a walkway should be provided at a safe distance.

9.2.15.   Adequate clearance should be provided between the track and buildings, structures, stacks of materials and other objects.

Operation

9.2.16.   The operation of inclined rope railways should be governed by signals in conformity with the requirements of section 1.8.

9.2.17.   Before loading begins at loading points wagons should be braked with hand brakes, brake shoes or other effective means.

9.2.18.   As a wagon or a car leaves the bottom station the switch should be moved to open the way to the dead-end siding.

9.2.19.   No person should walk along the track of a rope railway.

9.2.20.   Rope railways should not be inspected or repaired while they are in operation.

9.2.21.   Unauthorised persons should not be carried on rope railways.

10. Road and similar transport

10.1. Roads and travelways

10.1.1. Suitable roadways should be built to all places on construction sites to which road vehicles have to go.

10.1.2. Parts of the site that are dangerous for road vehicles should be:

(a) fenced off; or

(b) signposted.

10.1.3. Danger signs for road vehicles should be clearly visible by day or night.

10.1.4. Truck or tractor roads used for construction operations should be constructed and maintained so as to be safe for traffic in accordance with requirements which should be laid down by the competent authority.

10.1.5. In particular, truck roads should have gradients, surface, width and curves suitable for the traffic that they will have to carry.

10.1.6. Adequate guard-rails or fenders should be provided on bridges and alongside precipices, ravines and other declivities.

10.1.7. Bridges, trestles and their approaches should be inspected at suitable intervals.

10.1.8. Icy or otherwise slippery stretches of truck roads should be strewn with sand or other suitable non-slip material, especially on gradients and curves.

10.1.9. At approaches to level crossings, truck roads should not slope steeply.

10.2. Tractor and motor truck construction

General provisions

10.2.1. Tractors and motor trucks should be of sufficiently robust construction to withstand the heaviest stresses to which they will be subjected.

10.2.2. Motor trucks should be equipped with a cab or canopy and a seat for the driver, adequate brakes, safe means of getting on and off, lights, signalling equipment and, if necessary to prevent danger, mudguards for wheels.

10.2.3. Tractors should be similarly equipped as stated in paragraph 10.2.2.

Cab

10.2.4. The cab should be of such strength and be so installed as to provide adequate protection for the driver:

(a) if he is liable to be struck by falling or flying objects; or
(b) if the load is displaced.

10.2.5. The cab should be so arranged that:

(a) the interior is adequately ventilated and, if necessary, heated;
(b) the driver has an adequate field of vision.

10.2.6. The cab should be provided with:

(a) a windscreen and windows of transparent material that will not be broken into sharp splinters by an impact; and
(b) a power-driven windscreen wiper.

10.2.7. Tractors and motor trucks should be equipped with a footboard or steps and handholds such that it is possible to get into and out of the cab safely.

10.2.8. The cab should be so arranged that the drivers can easily get off the tractor or motor truck in an emergency.

10.2.9. If tractors or motor trucks operate on frozen lakes or watercourses, the cab should be so designed that the roof can be readily opened or removed from inside the cab. The roof should be open while the tractor is operated on frozen areas of water.

10.2.10. In the cab there should be a fixed seat for the driver and any passenger that the tractor or motor truck is allowed to carry.

10.2.11.   The driver's seat should:

(a) be so designed as to absorb vibrations sufficiently;

(b) have a back rest and a foot rest; and

(c) be generally comfortable.

10.2.12.   Control pedals should:

(a) be sufficiently wide;

(b) afford a secure foothold; and

(c) if necessary, be perforated so as to keep the surface clear of earth, mud, etc.

Brakes

10.2.13.   Tractors and motor trucks should be equipped with brakes that will hold them under the heaviest load that they may haul in any operating conditions and on any gradient for which the vehicle was designed.

10.2.14.   It should be possible to lock the brakes when the tractor or motor truck is stationary.

Exhaust pipes

10.2.15.   Tractor exhaust pipes should be:

(a) so placed as to prevent the build-up of harmful gases and fumes around the driver; and

(b) equipped with a spark arrestor.

Draw gear

10.2.16.   Motor trucks hauling trailers, and tractors, should be equipped with draw gear such that during coupling:

(a) no worker has to come between the vehicles being coupled while one is moving; or

(b) the vehicles being coupled cannot run into each other.

10.2.17.   Draw gear, including coupling pins, should be strong enough to hold the heaviest load that the tractor or truck may haul on any gradient and in any operating conditions.

10.2.18.   Coupling pins should be such that they cannot be accidentally lifted out of the coupling; if necessary, a span chain should be provided.

Hitching point

10.2.19.   Trailers and other equipment should be hitched to tractors in accordance with the manufacturer's instructions.

Lights

10.2.20.   Tractor and motor truck lights should comply with public traffic regulations even when they are used elsewhere than on public roads.

Starting mechanism

10.2.21.   Tractors and motor trucks should be provided with self-starting equipment.

10.2.22.   If a tractor or a motor truck has a hand crank, it should be secured against kicking back.

10.2.23.   Starting should be controlled by rotary or pull-out switches and not tumbler switches so as to reduce the risk of accidental starting.

Other equipment

10.2.24.   Tractors and motor trucks should be equipped with:

(a) a first-aid box or kit; and
(b) a suitable fire extinguisher.

10.2.25.   Tractors and motor trucks should be equipped with suitable audible signalling equipment.

Motor trucks with cranes

10.2.26.   Officially recognised stability tests should be established for motor trucks with cranes mounted.

10.3. Trailer construction

10.3.1.   Trailers equipped with steering devices should have an adequate seat or platform for the steerer.

10.3.2.   Steered trailers should be connected to motor trucks by an adequate signalling system.

10.3.3.   Trailers should be coupled to motor trucks by devices of adequate strength that can be securely fastened and locked.

10.3.4.   Trailer drawbars should be provided with substantial chains, straps or the like by which they can be safely lifted for coupling and uncoupling.

10.3.5.   When necessary, trailers should be equipped with jacks or similar supports to prevent tipping during loading.

10.4. Passenger vehicle construction

10.4.1.   Vehicles for transporting workers should comply with the relevant provisions of section 10.2 and with the provisions of this section.

10.4.2.   Vehicles used for transporting workers should have fixed seats for all the workers transported.

10.4.3.   If necessary to prevent danger, vehicles used for transporting workers should have:

(a) a cover for protection against the weather;
(b) a ladder or steps for the workers to climb on and off;
(c) enclosed racks or boxes for tools;
(d) a heating system; and
(e) an emergency light.

10.4.4.   Enclosed vehicles for transporting workers should have:

(a) an emergency exit as far away as practicable from the regular exit;
(b) effective means of signalling between the passengers and the driver;
(c) a lighting installation; and
(d) means of ventilation.

10.4.5. Vehicles regularly used for transporting workers should comply with official regulations for public passenger vehicles.

10.5. Inspection and maintenance of tractors and motor trucks

10.5.1. Tractors and motor trucks should be maintained in good running order, particular attention being paid to brakes and steering gear.

10.5.2. The cab, pedals, foot rests and steering gear should be kept clean.

10.5.3. Tractor and motor truck brakes should be tested at frequent intervals and, when necessary, adjusted.

10.5.4. Brakes, tyres, lights, steering gear, mirrors and wind-shield wipers should be inspected every day.

10.6. Operation of tractors and motor trucks

General provisions

10.6.1. Where necessary to prevent danger because of density of traffic, narrowness of the road, obstruction of the view or other reasons, a traffic control system should be operated.

10.6.2. Only competent persons should drive tractors and motor trucks.

10.6.3. Drivers of tractors and motor vehicles should wear adequate footwear and clothing.

10.6.4. Tractors and motor trucks should only be operated at places where conditions (as regards gradient, surface, obstructions, etc.) are such that they can be operated safely.

10.6.5. In particular, tractors and motor trucks should not be operated or started up in buildings if:

(a) there is a fire risk; or
(b) the ventilation is not adequate to prevent dangerous contamina-
    tion of the atmosphere.

10.6.6. The speed of tractors and motor trucks should be adapted to the operating conditions.

10.6.7. Before descending a steep slope the operator should engage an appropriate gear.

10.6.8. Tractors and motor truck drivers should keep a good look out for obstacles in their path.

10.6.9. Tractors and motor trucks should not haul loads too heavy for them to control on any sloping, uneven, soft or otherwise unsafe ground on which they will have to travel.

10.6.10. Tractors and motor trucks should not haul heavy vehicles or machines downhill unless the machines or vehicles can be adequately braked.

10.6.11. Tractors and motor trucks should be driven with particular care:

(a) over sloping, uneven, soft, slippery or otherwise unsafe ground;
(b) alongside ditches or banks;
(c) when turning;
(d) when reversing; and
(e) when hauling any trailer that considerably changes their centre of gravity.

10.6.12. Tractors should not push trucks, machines, etc. unless an adequate and securely fastened push bar is used for the purpose.

10.6.13. No person should get off a tractor or a motor truck except in emergencies unless:

(a) it is stationary; and
(b) there is a safe landing-place.

10.6.14. While a tractor or a motor truck is moving no person should:

(a) stand or sit in an unsafe place such as a roof, trailer, bar, mud-guard, running board, fender, or load;
(b) climb from one vehicle to another;
(c) apply wheel skids; or
(d) leave arms or legs projecting outside.

10.6.15. Motor trucks carrying passengers should not transport flammable liquids in bulk.

10.6.16. The transport of explosives should comply with the requirements of section 22.2.

10.6.17. Tractors should not carry:

(a) children; or

(b) loose objects, unless a safe place is provided for them.

10.6.18. Drivers of tractors and motor trucks should not drive for long periods without adequate rest.

10.6.19. When arriving at railway crossings, drivers of tractors and motor trucks should take great care, bringing their vehicle to a full stop if necessary, in any case extreme caution should be observed.

10.6.20. At places where men are working and at places where the driver does not have a clear field of vision, the movement of tractors and motor trucks should be controlled by signals.

10.6.21. Tractors and motor trucks should be driven so as to comply with official traffic regulations.

10.6.22. When coupling vehicles:

(a) if the motor vehicle is backed, the trailer should be blocked by the brake or chocks;

(b) if a trailer is pulled on to the motor vehicle, the trailer should, if necessary to prevent danger, be kept under control by the brake or chocks; and

(c) no person should remain between the motor vehicle and the trailer and, if practicable, the draw-bar should be handled with a hook or other suitable device.

10.6.23. When uncoupling vehicles, both vehicles should be blocked by brakes or chocks.

10.6.24. Vehicles should not be hitched to a tractor except by means of a draw-bar, at or below draw-bar level.

10.6.25. Tractors with winches should be properly aligned in the direction of the pull when pulling.

10.6.26. When filling the fuel tank the provisions of paragraph 12.2.5 should be complied with.

10.6.27. Radiator lids should be removed in such a way as to avoid the danger of scalding from steam or boiling water.

10.6.28. No person should go under a tractor or a motor truck without first notifying the driver, or making certain that the tractor or truck cannot move.

10.6.29. When leaving a tractor or a motor truck the driver should:

(a) put the engine out of gear;
(b) re-engage the master clutch, except in the case of torque converter machines;
(c) apply the brakes;
(d) lower any upraised equipment to a safe position;
(e) put power-operated attachments into the off position; and
(f) if necessary to prevent danger, block the wheels.

10.6.30. Workers should not be transported in dump trucks or trailers unless these are equiped with suitable safety devices and provided with adequate seating complying with section 10.4.

Loading and unloading vehicles

10.6.31. Vehicles being loaded or unloaded should be effectively braked or blocked.

10.6.32. Loads should be so stacked on vehicles that:

(a) the vehicle is not overloaded;
(b) the stability of the vehicle is not impaired;
(c) loads or parts of loads cannot cause danger to persons by projecting, shifting or falling; and
(d) loads or parts of loads cannot interfere with the safe driving or operation of the vehicle.

10.6.33. No load should be lowered on to a vehicle while any person is on the vehicle or in dangerous proximity to it.

10.6.34. Before a loaded motor truck starts on a journey, the load should be inspected to ensure that it is secure, evenly distributed and of a safe height, length and weight.

10.7. Traffic on ice

10.7.1. Motor vehicles should not cross frozen waterways unless the ice has a minimum thickness in accordance with the following table, and the area has been thoroughly inspected and tested for air pockets, slush holes and ice in poor condition:

Laden weight	*Minimum thickness of ice*
Trucks	
Up to 10 tons	45 cm (1 ft 6 in)
10-20 tons	60 cm (2 ft)
20-30 tons	75 cm (2 ft 6 in)
Tractors	
Up to 12.5 tons	50 cm (1 ft 8 in)

10.7.2. Crossings should:

(a) be at intervals of not less than 125 m (420 ft); and

(b) be cleared of snow to a width of at least 17.5 m (60 ft).

10.7.3. Only one-way traffic should be allowed on crossings.

10.7.4. Intervals between motor vehicles on crossings should be at least 40 m (135 ft).

10.7.5. Cab doors of trucks should be secured open when they are travelling on ice.

10.8. Traffic on roads under construction

10.8.1. Workers employed on road construction work should be protected from traffic by barricades, signs, lights, watchmen or other effective means.

10.8.2. Signs and lights should be placed where:

(a) they can easily be seen by approaching traffic; and

(b) they will not be obscured by splashes from traffic.

10.8.3. Signs and lights should not be obscured by equipment, stacks of material, etc.

10.8.4. At night, barricades, signs and hazardous conditions should be marked by lights or reflectors.

10.8.5. As far as practicable, a separate traffic lane should be provided for construction equipment.

10.8.6. Where construction equipment has to use or cross public traffic lanes:

(a) a traffic control system should be operated; or
(b) a watchman should direct the movements of the equipment.

10.8.7. Road construction equipment should be painted a distinctive conspicuous colour such as yellow.

10.9. Power buggies

10.9.1. Power buggies should only be operated by suitably trained persons.

10.9.2. Power buggies should be designed to turn sharply, stop suddenly and operate on slopes under the full load and maximum speed without tilting.

10.9.3. Power buggies should be equipped with brakes capable of stopping them quickly under the full load and maximum speed.

10.9.4. The controls of power buggies should be so arranged and protected that they cannot be accidentally actuated.

10.9.5. Power buggies should be equipped with an adequate horn or other audible warning device.

10.9.6. Power buggies on which the operator rides should be equipped with a safe seat or platform.

10.9.7. The speed of power buggies in which the operator rides should be limited.

10.9.8. Power buggies should not be left unattended on a slope on which they could run away.

10.10. Lift and other industrial trucks

Construction

10.10.1. The capacity rating should be clearly incised or stamped on every lift and fork-lift truck.

10.10.2. Fork-lift trucks should be equipped with an overhead canopy or guard adequate to prevent injury to the operator from falling objects.

10.10.3. All industrial trucks should be equipped with a horn, gong, whistle or other adequate warning device.

10.10.4. Every power-operated lift and fork-lift truck should be equipped with a lockable brake.

10.10.5. It should be possible to lock the elevating mechanism of lift and fork-lift trucks in any position.

10.10.6. Pallet trucks with hand-lift handles should be equipped with an automatic device to retain the raised load and free the handle until it is re-engaged by the operator.

10.10.7. Fork-lift trucks handling small objects or unstable loads should be equipped with an adequate load backrest to prevent any part of the load from falling towards the mast.

10.10.8. Fork extensions should be adequately secured against displacement.

10.10.9. Electric trucks on which the driver rides should be equipped with a switch that automatically opens the driving circuit when the driver leaves the truck.

10.10.10. Pedals and operating platforms on trucks on which the driver stands should have a non-slip surface.

Operation

10.10.11. Lift and other industrial trucks should be operated by suitably trained persons.

10.10.12. While lift and fork-lift trucks are moving, the load should be kept as low as practicable.

10.10.13.   Lift and fork-lift trucks should not be used on dangerously uneven surfaces.

10.10.14.   Fork-lift trucks should not be loaded or unloaded while they are moving.

10.10.15.   If the load obscures the driver's view the operation of a fork-lift truck should be governed by signals.

10.10.16.   No person other than the driver should ride on a powered industrial truck unless there is a safe stand or seat for him.

10.10.17.   Pallets should:

(a) not be supported on unstable material; and

(b) be kept level.

10.10.18.   No packaged units bound by means of wire or metal tape should be handled by a lift or a fork-lift truck if any part of the wire or tape is broken.

10.10.19.   Precautions should be taken to prevent spillage of the load.

10.11. Animal transport

10.11.1.   Animal-drawn vehicles should be provided with adequate brakes.

10.11.2.   Vehicles on which drivers or brakesmen have to travel should be provided with a safe seat for each person travelling.

10.11.3.   Vehicles transporting workers should have a safe seat for every worker.

10.11.4.   Single-axle vehicles should, if necessary to prevent danger, be secured against tipping up.

10.11.5.   Draught animals should be treated kindly and handled gently.

10.11.6.   When animals are being led, the lead should not be wrapped round the wrist or waist.

10.11.7.   Care should be taken that the harness is comfortable and, in particular, does not abrade or tear the animal's skin.

10.11.8.   Animals that bite should be muzzled when not in the stall.

10.11.9.   Animals that kick should only be handled by persons familiar with them.

10.12. Wheelbarrows

10.12.1.   Wheels of wheelbarrows should run true and be well secured to the frame.

10.12.2.   Wheelbarrows with split or cracked handles should not be used.

10.12.3.   When wheelbarrows are not in use they should be so left that they cannot easily tip over.

11. Garages

11.1. General provisions

11.1.1. Garages for motor vehicles should:

(a) be of fire-resistant construction;
(b) be adequately ventilated; and
(c) have at least one outside wall.

11.1.2. Electrical installations in garages should comply with the relevant requirements of Chapter 17.

11.1.3. Heating installations in garages should have no open flames or incandescent parts.

11.1.4. Inspection pits in garages should be:

(a) accessible by means of safe steps; and
(b) when not in use, provided with safe covers or perimeter barricades or guard-rails or toe-boards.

11.1.5. Garage floors should have a drainage system such that:

(a) there is a trap for petrol (gasoline) and oil; and
(b) the trap can be emptied easily.

11.1.6. Drain traps should be emptied at suitable intervals.

11.1.7. Large quantities of fuel and oil should not be stored in garages.

11.1.8. Special precautions should be taken when work that involves welding or the generation of sparks is carried out in garages; in particular the relevant requirements of Chapter 28 should be complied with.

11.1.9. Oily and greasy waste should be kept in a self-closing metal receptacle.

11.1.10. At an easily accessible place in a garage there should be kept in readiness for use:

(a) suitable fire-extinguishing equipment; or
(b) an adequate quantity of dry sand and a shovel.

12. Engines

12.1. General provisions

12.1.1. Engines should:

(a) be so constructed and installed that they can be started safely;
(b) be so constructed and installed that the maximum safe speed cannot be exceeded;
(c) have remote controls for limiting speed; and
(d) have devices that will enable them to be stopped from a safe place in emergencies.

12.1.2. Engines should be operated by competent persons.

12.1.3. On engines, moving parts such as flywheels, driving belt pulleys, cranks and crossheads should be enclosed or adequately guarded.

12.2. Internal combustion engines

12.2.1. Starting cranks of internal combustion engines should have double bearings or be otherwise effectively secured against kicking back.

12.2.2. Internal combustion engines should not be started up with oxygen or combustible gases.

12.2.3. Exhaust gases from internal combustion engines should be so led off that the attendant and other persons in the vicinity are not exposed to them.

12.2.4. Internal combustion engines, such as those of tractors and motor vehicles, should not run for long periods of time in closed premises unless adequate mechanical ventilation is provided or gases are directly exhausted outside or the equipment is provided with a fume scrubber.

12.2.5. When internal combustion engines are being fuelled:
(a) suitable equipment such as pumps, hoses and nozzles should be used;

(b) the engine ignition should be shut off;

(c) care should be taken to avoid spilling fuel;

(d) no person should smoke or have an open light in the vicinity; and

(e) a fire extinguisher should be kept readily available.

12.3. Stationary donkey engines

12.3.1. Stationary donkey engines should be installed on a firm, level foundation and securely anchored in position.

12.3.2. Stationary donkey engines should be provided with an adequate running board on each side.

12.3.3. Where necessary to prevent danger from broken ropes, donkey engines should be provided with a substantial roof.

12.2.4. Walkways alongside donkey engines should be provided with adequate guard-rails and toe-boards.

12.3.5. Donkey engines should be equipped with an acoustic signalling device.

12.3.6. Regulations should be established for donkey-engine boilers concerning their construction, installation, operation, maintenance, testing and examination.

12.3.7. In particular, all boilers should:

(a) be provided with one or more safety valves, a water-level gauge, a steam pressure gauge, gauge cocks and a blow-off pipe; and

(b) be tested by a competent person or authority before being taken into use.

12.3.8. Blow-off and exhaust pipes should be so situated or guarded that no person can be endangered by accidental contact with them or by escaping steam or hot water.

12.3.9. Gauge glasses should be equipped with a substantial wire mesh or equivalent guard.

12.4. Tractor power take-offs

12.4.1. Tractor power take-offs should be so guarded that while the engine is running:

(a) if the power take-off is in use, it is covered on top and at both sides by a shield attached to the tractor that prevents any person from coming into contact with the power take-off; and

(b) if the power take-off is not in use, it is completely enclosed in a cover attached to the tractor.

12.4.2. Power take-off shields and covers should be capable of supporting a weight of 110 kg (250 lb) when attached to the tractor.

12.4.3. Power take-off shafts, including their universal joints, should, while in motion, be completely enclosed in a guard that prevents any person from coming into contact with the shaft.

12.4.4. Protective devices for power take-offs and power take-off shafts should be:

(a) substantially constructed;

(b) firmly secured in position; and

(c) maintained in good condition.

13. Machinery, general provisions

13.1. Construction and installation

13.1.1. All dangerous parts of moving machinery, including the point of operation, should be effectively guarded unless they are so constructed, installed or placed as to be as safe as if they were guarded by appropriate safety devices.

13.1.2. To ensure the safety referred to in paragraph 13.1.1, in particular:

(a) all flywheels, gearing, cone and cylinder friction drives, cams, pulleys, belts, chains, pinions, worm gears, crank arms and slide blocks and, to the extent prescribed by the competent authority, shafting (including the journal ends) and other transmission machinery also liable to present danger to any person coming into contact with them when they are in motion, should be so designed or protected as to prevent such danger; controls should also be so designed or protected as to prevent danger;

(b) all set-screws, bolts and keys and, to the extent prescribed by the competent authority, other projecting parts of any moving part of machinery also liable to present danger to any person coming into contact with them when they are in motion should be so designed, sunk or protected as to prevent such danger;

(c) paths of counterweights, pendulum weights and the like should be fenced; and

(d) all working parts of machinery which, while in operation, may produce flying particles should be, as far as is practicable, adequately guarded.

13.1.3. Every power-driven machine should be provided with adequate means, immediately accessible and readily identifiable to the operator, of stopping it quickly, and preventing it from being started again.

13.1.4. Machine operators' stands should:

(a) be safely and easily accessible;

(b) be sufficiently spacious; and

(c) be so designed and constructed as to permit safe operation of the machine and not to cause the operator undue fatigue or discomfort.

13.1.5. If necessary to prevent danger, machine operators' stands should be provided:

(a) with fencing, guard-rails and toe-boards or the like; and

(b) with special means of access such as steps and/or grips.

13.1.6. If the operator's stand is totally enclosed, adequate means of heating and/or ventilation should be provided if necessary.

13.1.7. Machine controls should be so designed and installed that they can be operated reliably, safely and easily from the operator's stand.

13.1.8. On power-driven machines the maximum safe speed and, if appropriate, the direction of revolution should be indicated.

13.1.9. Machinery should be provided with the requisite protective devices even when it is idle for a considerable time, unless it has been rendered inoperative.

13.1.10. Openings in guards should not exceed the following dimensions according to the distance between the guard and the machine part that it is protecting:

(a) 6 mm (¼ in) when the distance is less than 10 cm (4 in);

(b) 1.2 cm (½ in) when the distance is between 10 and 40 cm (4 in and 1 ft 4 in);

(c) 5 cm (2 in) when the distance is more than 40 cm (1 ft 4 in).

13.1.11. Pits and floor openings for moving parts should be protected by guard-rails and toe-boards complying with the requirements of paragraphs 2.6.1 to 2.6.5.

13.1.12. Over workplaces and passageways, fast-running belts, ropes and steel bands, and also large belts, should be provided with a catch platform underneath.

13.1.13. When machines are being installed, repaired or transported, precautions should be taken to prevent them from accidentally starting up.

13.1.14. If necessary to prevent danger, certain parts of compressors, electrical installations and other equipment should be adequately protected against the weather and mechanical damage.

13.2. Inspection, maintenance of machinery

13.2.1. Machinery operators' stands should be properly maintained and be kept clear of obstructions.

13.2.2. The fencing of dangerous parts of machinery should not be removed while the machinery is in use, but if removed should be replaced as soon as practicable, and in any case before the machinery is taken into normal use again.

13.2.3. No part of any machinery which is in motion and which is not securely fenced should be examined, lubricated, adjusted or repaired except by duly authorised persons, in conformity with accepted standards of safety.

13.2.4. Machine parts should only be cleaned when the machine is stopped, unless cleaning can be carried out in conformity with accepted standards of safety.

13.2.5. Belts, ropes, chains and cords of transmission should not be mounted or dismounted by hand while the machinery is in motion.

13.2.6. When machinery is stopped for servicing or repairs, adequate measures should be taken to ensure that it cannot be inadvertently restarted, without clearance by the maintenance crew.

13.2.7. When repair, maintenance or other work has to be done in dangerous proximity to machinery, the machinery should be stopped for the duration of the work.

13.3. Operation of machinery

13.3.1.   Only reliable and competent persons who have been given adequate instruction should be entrusted with the independent operation of transmissions and working machines.

13.3.2.   Machinery operators should:

(a) not wear loose clothing, ties, scarves or jewellery; and
(b) cover loose hair that might be caught by moving parts.

13.3.3.   Before they are started, machines should be examined to ensure that they are in safe working condition, and in particular that:

(a) all adjustments have been made;
(b) working parts are properly lubricated;
(c) nuts and bolts are properly tightened; and
(d) all protective devices are in place and properly secured.

13.3.4.   Machinery in motion should not be left unattended if this could cause danger.

13.3.5.   If danger could arise when any machinery is started, immediately before starting, a signal should be given that can clearly be heard or seen at the place where the machinery is installed.

13.3.6.   If a number of persons are employed simultaneously at a machine that can be started by a prime mover, the machine should not be started until the person starting it has satisfied himself that no person is in danger.

13.3.7.   Transmissions and working machines that can be disengaged should only be disengaged when they are stopped, and should remain disengaged while they are idle.

13.3.8.   Adequate measures should be taken to prevent:

(a) the maximum safe speed from being exceeded; and
(b) sudden changes in speed.

13.3.9.   Machines made for manual power operation should not be power driven.

13.3.10. If in the use of machines danger could be caused by flying sparks, splinters, chips, dust or the like, adequate measures should be taken to eliminate the danger.

13.3.11. In particular, adequate measures should be taken to prevent eye injuries.

13.3.12. Transmission belts should not be thrown on or off, or adjusted, while in motion.

13.3.13. Machinery which continues to run after power has been disconnected, and may thus introduce a risk, should be provided with a brake, which should be capable of being operated from the control point.

14. Woodworking machines

14.1. General provisions

14.1.1. Woodworking machines should only be operated by competent persons.

14.1.2. Operators of woodworking machines should not be disturbed while the machine is working.

14.1.3. Operators of non-automatic woodworking machines should not leave them without stopping them or enclosing the tools.

14.1.4. Woodworking machines should not be adjusted or cleared of jammed wood while they are working.

14.1.5. Shavings, sawdust, etc. should not be removed by hand from woodworking machines or in their vicinity, while the machines are working.

14.1.6. Woodworking machines that use tools of widely differing diameters should have a device for altering the speed of rotation.

14.1.7. If the speed of woodworking machines can be varied:

(a) it should only be possible to start them at the lowest speed; and

(b) the operating speed should be indicated.

14.1.8. Workpieces should be firmly supported, or be securely guided or clamped.

14.1.9. Free ends of long workpieces should be supported on trestles, table extensions, etc.

14.1.10. Small or short workpieces should be guided, clamped or pushed with a push stick.

14.2. Circular saws

Construction

14.2.1. Circular saws should be provided with hood guards.

14.2.2. Hood guards should:

(a) cover as far as practicable all the exposed part of the saw above the table;

(b) be easily adjustable; and

(c) protect the operator from accidental contact with the saw, and from flying splinters or broken saw teeth.

14.2.3. Parts of circular saws underneath tables should be closely covered with guards or hoods.

14.2.4. Circular saws should be provided with strong, rigid and easily adjustable riving knives of suitable design.

14.2.5. The width of the opening in the table for the saw blade should be as small as practicable.

14.2.6. Carriages or roll tables should be secured against jumping or running off the rails.

14.2.7. Portable circular saws should be so designed that when the blade is running idle it is automatically covered.

14.2.8. It should not be possible to block the cover of a portable circular saw.

14.2.9. Where necessary a suitable push stick should be used with every hand-fed circular saw.

Inspection, maintenance

14.2.10. Circular saws should:

(a) be properly maintained, set and sharpened;

(b) examined at frequent intervals; and

(c) replaced or removed for correction if defects are discovered.

Operation

14.2.11. The maximum speed of circular saws should not exceed that recommended by the manufacturer.

14.2.12.   Workers should not adjust saw blades or guides while the saw is running if this would cause danger.

14.2.13.   Saw blades should not be braked by pressing on them after the power has been cut off.

14.2.14.   When round timber, poles and the like are being cross cut, devices should be used to prevent them from twisting or tipping.

14.2.15.   Precautions should be taken to prevent small pieces cut off from being caught by the saw blade.

14.2.16.   Precautions should be taken to prevent the work from kicking back.

14.3. Band saws

Construction

14.3.1.   On band saws all the blade should be covered up to the cutting point.

14.3.2.   Band wheels on band saws should be enclosed with stout guards of sheet metal or other material of at least equivalent strength.

14.3.3.   Guards for upper band wheels should extend:

(a) downwards to below the lower part of the wheel rim; and
(b) upwards so as to afford a clearance at the top of not less than 10 cm (4 in) from the wheel.

14.3.4.   Guards for lower band wheels should:

(a) serve as a means of protection for the area beneath the saw table; and
(b) permit the removal of sawdust from the area beneath the table so that the saw blade may run free at all times.

14.3.5.   The working side of the blade between the guide rolls or gauges and the upper wheel enclosures should also be enclosed by a protector which as far as practicable should be of a self-adjusting type.

14.3.6. Band saws should be provided with an automatic tension regulator.

Inspection, maintenance

14.3.7. Band saws should:

(a) be properly maintained, set and sharpened;

(b) examined at frequent intervals; and

(c) replaced or removed for correction if defects are discovered.

14.3.8. Workers should not attempt to remove broken blades while the machine is running.

14.3.9. If the blade guide is operated by hand the machine should be stopped before it is adjusted.

14.3.10. When round timber, poles or the like are being sawn, devices should be used to prevent them from twisting or tipping.

14.4. Planing machines

Construction

14.4.1. Only cylindrical cutter blocks should be used on overhand planing machines.

14.4.2. Overhand planing machines should be provided with bridge guards capable of covering the full length and breadth of the cutting slot in the bench, and so constructed as to be easily adjusted in both a horizontal and a vertical direction.

14.4.3. Table openings on overhand planing machines should be as small as practicable.

14.4.4. Cutter knives exposed under the table should be guarded.

14.4.5. The feed roller of thicknessing machines should be provided with an adequate guard.

14.4.6. Thicknessing machines should be provided with a kick-back preventer which should be kept as free as possible.

14.4.7. The kick-back preventer should be constructed of elements which:

(a) are not more than 1.5 cm (³/₅ in) apart;
(b) are secured against excessive swinging; and
(c) fall back automatically after raising.

Operation

14.4.8. When short pieces are being planed a feeding device should be used.

14.4.9. When wood is being grooved a pressure device should be used.

15. Construction equipment

15.1. Earth-moving equipment: general provisions

Construction

15.1.1. Earth-moving equipment should be provided with a plate or the like indicating:

(a) the gross laden weight;

(b) the maximum axle weight or, in the case of caterpillar equipment, ground pressure; and

(c) the tare weight.

15.1.2. Earth-moving equipment should be equipped with:

(a) an electrically operated acoustic signalling device;

(b) searchlights for forward and backward movement;

(c) power and hand brakes;

(d) tail lights; and

(e) silencers.

15.1.3. Operators of earth-moving equipment should be adequately protected against the weather by a cab, windscreen, roof or other effective means.

15.1.4. Cabs of earth-moving equipment should comply with the relevant requirements of section 10.2.

15.1.5. Earth-moving equipment with a cab should be equipped with:

(a) a direction indicator; and

(b) a rear-view mirror on either side.

15.1.6. Operator's seats of earth-moving equipment should comply with the relevant requirements of section 10.2.

15.1.7. Operators of equipment with drag lines or winch ropes should be adequately protected against being struck by broken ropes.

15.1.8.   If necessary to prevent danger, operators of dumping equipment should be protected against being struck by parts of the load.

Operation

15.1.9.   No earth-moving equipment should be started up until all workers are in the clear.

15.1.10.   The cab of earth-moving equipment should be kept at least 1 m (3 ft 3 in) from a face being excavated.

15.1.11.   When cranes and shovels are travelling, the boom should be in the direction of travel and the scoop or bucket should be raised.

15.1.12.   The buckets or scoops of shovels and cranes should not be loaded when the equipment is travelling.

15.1.13.   Earth-moving equipment should not travel on bridges, viaducts, embankments, etc., unless it has been found safe for it to do so.

15.1.14.   Bulky material such as beams, girders and poles should not be carried in the scoops of earth-moving equipment, unless the machines are specially designed for the purpose.

15.1.15.   No person should enter the radius of action of earth-moving equipment when in operation.

15.1.16.   Adequate precautions should be taken to prevent earth-moving equipment being operated in dangerous proximity to live electrical conductors.

15.1.17.   On earth-moving equipment motors, brakes, steering gear, chassis, blades, blade-holders, tracks, wire ropes, sheaves, hydraulic mechanisms, transmissions, bolts and other parts on which safety depends should be inspected daily.

15.1.18.   Dusty haulage roads and tracks should be watered so as to maintain good visibility.

15.1.19.   Earth-moving equipment should not be left on a slope with the engine running.

15.1.20.   As far as practicable earth-moving equipment should not be left on highways at night.

15.1.21.   If earth-moving equipment is left on a highway it should be adequately marked with lanterns, red flags or other effective means.

15.1.22.   Unauthorised persons should not be conveyed on earth-moving equipment.

15.1.23.   No adjustments, maintenance work or repairs should be made on equipment in motion.

15.1.24.   Deck plates and steps should be kept free from oil, grease, mud or other slippery substances.

15.1.25.   Dredge-type excavators should not be used on earth walls more than 1 m (3 ft 3 in) higher than the reach of the excavator if they are installed at the bottom of the wall.

15.1.26.   Bucket excavators should not be used at the top or bottom of earth walls with a slope exceeding 60°.

15.2. Power shovels, excavators

General provisions

15.2.1.   Power shovels (excavators) should be operated in such a way that they do not lose stability.

15.2.2.   Tracks for track-mounted power shovels should comply with the relevant requirements of section 5.4.

15.2.3.   If necessary to prevent danger during inspection or repair, the jib of power shovels should be equipped with a ladder protected by a guard-rail and toe-board.

15.2.4.   Brake pedals for all motions on power shovels should have two independent locking devices.

15.2.5.   Power shovels should be equipped with an emergency quick-acting stop device independent of the controls.

15.2.6.   Excavators that are equipped with a unit for deep digging should either be so designed that the bucket teeth cannot come nearer the boom than 40 cm (1 ft 4 in) or be provided with a reliable stop that prevents this from happening.

15.2.7. Excavators that are used for lifting with lifting gear should be provided with a plate on the boom bearing a clearly legible and durable text giving the maximum permissible load on the lifting gear fitted.

15.2.8. The maximum load referred to in paragraph 15.2.7 should apply to the most unfavourable position as regards stability in which the excavator can be used for lifting when it stands on a horizontal firm base.

15.2.9. Excavators that are equipped for use as mobile cranes should be inspected and tested as required for lifting appliances.

Steam shovels

15.2.10. Official regulations concerning the construction, installation, operation, testing and examination of steam boilers for steam shovels should be established.

15.2.11. In particular boilers should:

(a) be provided with one or more safety valves, a water-level gauge, a steam pressure gauge, gauge cocks and a blow-off pipe; and

(b) be tested by a competent person or a competent authority before being taken into use.

15.2.12. Blow-off and exhaust pipes should be so situated or guarded that no person can be endangered by accidental contact with them or by escaping steam or hot water.

15.2.13. Boilers and steam pipes should be insulated so as to protect workers against the risk of burns.

Petrol-operated shovels

15.2.14. Petrol-operated shovels should be:

(a) earthed or otherwise protected against static electricity; and
(b) equipped with a fire extinguisher.

Electric shovels

15.2.15. The connection or disconnection of the electric cable supplying power from the transmission line to or from the electric shovel should only be done by competent persons duly authorised.

15.2.16. Electrical operating connectors and relays on the shovel should be inspected daily if in operation.

Operation of shovels (excavators)

15.2.17. Excavator operators should be:

(a) at least 18 years of age; and
(b) familiar with the operation and maintenance of the machine.

15.2.18. Power shovels should be so placed that:

(a) there is ample space for the operations;
(b) the operator has a clear view over the area of operation; and
(c) there is no danger of tipping, slipping or overturning.

15.2.19. While a power shovel is working:

(a) no one should enter the area of operation without first warning the operator; and
(b) no one should work, pass or stand under the raised bucket or grab.

15.2.20. No unauthorised person should be allowed on the operating platform while the shovel is in operation.

15.2.21. The boom should be prevented from accidentally swinging during operations or transport.

15.2.22. The bucket or grab of a power shovel should be prevented from accidentally dipping, tipping or swinging in operation.

15.2.23. Before leaving the shovel the operator should:

(a) disengage the master clutch; and
(b) lower the bucket or grab to the ground.

15.2.24. Buckets or grabs of power shovels should be fixed to restrict movement while they are being repaired or teeth are being changed.

15.2.25. When an excavator is at work near a wall or similar construction, persons should be prevented from entering the danger zone in which they may be crushed when the machine pivots.

15.2.26. Trucks should not be loaded in any place where there may be danger from materials such as rocks falling from buckets passing overhead; where this cannot be avoided, no person should remain in the cab during loading.

15.2.27. Trucks should be stationed at such a distance from the excavator that there is a clearance of at least 60 cm (2 ft) between the truck and the superstructure of the excavator even when it pivots.

15.2.28. While work is being done on hydraulically operated buckets the piston should be fully drawn back in the hydraulic cylinder.

15.3. Bulldozers

15.3.1. Before leaving a bulldozer the operator should:

(a) apply the brakes;
(b) lower the blade; and
(c) put the shift lever in neutral.

15.3.2. At the close of work bulldozers should be left on level ground.

15.3.3. When a bulldozer is moving uphill the blade should be kept low.

15.3.4. Bulldozer blades should not be used as brakes except in an emergency.

15.4. Scrapers

15.4.1.   The tractor and scrapers should be joined by a safety line when in operation.

15.4.2.   Scraper bowls should be blocked while blades are being replaced.

15.4.3.   Scrapers moving downhill should be left in gear.

15.5. Asphalt plants

General provisions

15.5.1.   Asphalt plants should be provided with:

(a) safe means of access and working platforms; and
(b) suitable fire-extinguishing appliances and materials.

15.5.2.   Elevated platforms on asphalt spreaders should:

(a) be protected by railings complying with the requirements of paragraphs 2.6.1 to 2.6.5; and
(b) be provided with an access ladder complying with the relevant requirements of Chapter 4.

15.5.3.   Wooden floors in front of the sprayers should be covered with corrugated sheet metal.

15.5.4.   The mixer elevator should be enclosed in a wooden or sheet-metal shaft.

15.5.5.   The shaft referred to in paragraph 15.5.4 should have a window for observation, lubrication and maintenance.

15.5.6.   Bitumen scoops should have solid covers.

15.5.7.   Where necessary mixers should have the top adequately covered by netting.

15.5.8.   The sprayer should be provided with a fire-resisting shield.

15.5.9.   The shield referred to in paragraph 15.5.8 should have an observation window.

15.5.10.   Piping for hot oil and asphalt should be adequately insulated to protect workers from injury by burns.

15.5.11.   Flexible piping working under positive pressure should be metal encased.

15.5.12.   To avoid fire risks due to foaming:

(a) boilers should have a device that prevents foam from reaching the burners; or
(b) only non-foaming products should be used.

15.5.13.   Where necessary, spreader tank cars and gravel spreaders should have a platform with railings for the operator.

15.5.14.   When necessary to prevent danger from rock dust in asphalt plants:

(a) the screen enclosure and the mixing chamber should be equipped with a mechanical exhaust system;
(b) the screen overflow chutes and hoppers should be enclosed;
(c) the waste conveyor system should be enclosed at transfer and discharge points;
(d) all material transfer points should be equipped with effective dust prevention devices;
(e) conveyor and elevator covers should be dust tight;
(f) the dryer discharge to the screen elevator should be equipped with an effective dust seal; and
(g) contaminated air should be so discharged that it cannot return to working areas.

Operation

15.5.15.   The operation of asphalt plants should be under the permanent control of a competent person.

15.5.16.   When asphalt plants are working on a public road an adequate traffic control system should be established.

15.5.17.   Ample storage space should be provided for materials, vehicles, etc., so that operations are not obstructed.

15.5.18. A sufficient number of fire extinguishers should be kept in readiness on the work site, including at least two on the spreader.

15.5.19. Workers handling hot asphalt should wear gloves, rubber boots, goggles and, if necessary to prevent danger, suitable protective clothing.

15.5.20. Burners should be lighted by butane lights or other suitable devices, and not by burning rags and the like.

15.5.21. When not in use flexible piping should not be left on the ground.

15.5.22. Heating tubes in boilers should be kept well covered with asphalt.

15.5.23. Material should only be loaded on to the elevator after the drying drum has warmed up.

15.5.24. No open light should be used for ascertaining the level of asphalt.

15.5.25. Thinners (cut-backs) should not be heated over an open flame.

15.5.26. If a burner goes out:

(a) the fuel supply should be cut off; and

(b) the heating tube should be thoroughly blown out by the fan, so as to prevent a backfire.

15.5.27. Piping should not be warmed with burning rags.

15.5.28. Spilled asphalt should be promptly cleaned up around boilers.

15.5.29. Inspection openings should not be opened while there is any pressure in the boiler.

15.5.30. The drying drum and the mixer should not be inspected or repaired while they are in motion.

15.5.31. When tanks are cleaned with steam, adequate precautions should be taken to prevent any build-up of pressure.

15.5.32. Spreaders in operation should be protected by signals, signs or other effective means.

15.5.33. Gravel spreaders, asphalt spreaders and rollers should always be kept at a safe distance from each other to avoid danger.

15.6. Pavers

15.6.1. Pavers should be equipped with a loud acoustic signalling device.

15.6.2. The signal should be sounded before:

(a) the paver moves forward; and

(b) the bucket is run out to workers.

15.6.3. Pavers should be equipped with guards that prevent workers from walking under the skip.

15.6.4. The operation of trucks working with pavers should be governed by signals in accordance with the requirements of section 1.8.

15.7. Road rollers

15.7.1. Before a road roller is used the ground should be examined for bearing capacity and general safety, especially at the edges of slopes such as embankments.

15.7.2. No person should climb on to a moving roller.

15.7.3. High-powered engines of rollers should not be started up by hand.

15.7.4. Rollers should not move downhill with the engine out of gear.

15.7.5. When a roller is not in use:

(a) the brakes should be applied;

(b) the engine should be put into bottom gear if the roller is facing uphill;

(c) the engine should be put into reverse if the roller is facing downhill;

(d) the contact should be switched off; and

(e) the wheels should be blocked.

15.7.6. As far as practicable rollers should not be left on a highway after the close of work.

15.8. Concrete mixers

General provisions

15.8.1. All gears, chains and rollers of concrete mixers should be adequately guarded to prevent accidental contact.

15.8.2. Concrete mixer skips should be protected by side railings to prevent workers from passing under them while they are raised.

15.8.3. If the operator's stand is more than 1.5 m (5 ft) above the ground, it should be provided with:

(a) safe means of access such as ladders complying with the relevant requirements of Chapter 4; and

(b) guard-rails and toe-boards complying with the requirements of paragraphs 2.6.1 to 2.6.5.

15.8.4. Hoppers into which a person could fall, and revolving blades of trough or batch type mixers, should be adequately guarded by grating.

15.8.5. In addition to the operating brake, skips of concrete mixers should be provided with a device or devices by which they can be securely blocked when raised.

Operation

15.8.6. No worker should go under the skip unless it has been securely blocked by two independent methods.

15.8.7. Concrete mixer operators should not lower a skip before making sure that all workers are in the clear.

15.8.8. The area round concrete mixers should be kept clear of obstruction.

15.8.9. While the drum of a concrete mixer is being cleaned adequate precautions should be taken to protect the workers inside, for instance by locking switches open, removing fuses or otherwise cutting off the power.

15.8.10. Ropes and sheaves of concrete mixers should be inspected on each day of operation.

15.9. Loading machines (belt or wheel loaders)

15.9.1. Loading machines should be equipped with a cab affording protection against impact.

15.9.2. If there is a risk of the operator's being injured by crushing between the bucket jib and the fixed parts of the machine the cab should comply with the requirements of paragraphs 15.9.3 to 15.9.8.

15.9.3. Side doors of the cab should be so arranged that no contact with the bucket jib is possible when the doors are open.

15.9.4. Hinged doors should be so hung that they cannot easily be taken off, for instance the hinge should be secured by a riveted bolt or other means.

15.9.5. Side windows that can be opened or removed, and other openings in the cab where the operator might be injured if he put his hand or arm out, should be covered by strong, sufficiently tight grating.

15.9.6. If the glass of a side window that has no grating is broken it should be immediately changed.

15.9.7. Roof hatches or back frames that can be opened should be arranged as emergency exits.

15.9.8. In the cab there should be a plate warning against removing the door, unprotected side frames or gratings.

16. Hand tools, portable power-driven tools

16.1. Hand tools

Materials and construction

16.1.1. Hand tools and implements should be of material of good quality and appropriate for the work for which they will be used.

16.1.2. Wooden handles of hand tools and implements should be of hard, straight-grained wood free from cracks and knots.

16.1.3. Handles of hand tools and implements should be fitted carefully to the heads, finished smooth, and kept securely fastened to them.

16.1.4. The handles of machetes and similar cutting tools should have a projection that prevents the hand from slipping on to the blade.

Maintenance

16.1.5. Hand tools and implements should be tempered, dressed and repaired by competent persons.

16.1.6. The cutting edges of cutting tools should be kept sharp.

16.1.7. Heads of hammers, wedges and other shock tools should be dressed or ground to a suitable radius on the edge as soon as they begin to mushroom or crack.

Transport

16.1.8. While being transported, the edges or points of sharp-edged or sharp-pointed hand tools such as axes should be so placed, buried or sheathed as to prevent danger.

16.1.9. Sharp-edged and sharp-pointed tools should not be carried on bicycles unless they are so protected by sheaths and secured in position as not to cause danger.

16.1.10.   Unless adequately protected, sharp-edged and sharp-pointed tools and implements and glass bottles should not be carried in pockets.

16.1.11.   Saws carried on the shoulder should have the teeth pointed outwards.

16.1.12.   When unsheathed axes are carried they should be gripped near the head with the arm extended and the blade parallel to the leg.

Storage

16.1.13.   When not in use sharp tools should be kept in sheaths, shields, chests or other suitable containers.

16.1.14.   Sharp-edged and sharp-pointed hand tools and implements should be so stored that:

(a) the edges and points are out of reach or are otherwise prevented from causing danger;

(b) they cannot fall; and

(c) they cannot cause danger to the person removing them.

Handling and use

16.1.15.   Hand tools and implements should be used only for the specific purposes for which they were designed.

16.1.16.   Sharp-edged and sharp-pointed tools and implements should not:

(a) be thrown from person to person;

(b) be used in dangerous proximity to other persons or moving machinery; or

(c) be used as props, rammers, prods or the like.

16.1.17.   Hand tools and implements should not be left lying in places where persons have to work or pass, or on scaffolds or other elevations from which they might fall on persons below.

16.1.18.   Only insulated or non-conducting tools should be used on or near live electrical installations if there is any risk of electrical shock.

16.1.19.   Only non-sparking tools should be used near flammable materials or in the presence of explosive dusts or vapours.

16.1.20.   Open-jawed wrenches should be placed on nuts with the jaw opening facing the direction in which the handle will move.

16.1.21.   Open-jawed wrenches should be pulled, not pushed.

16.1.22.   Wrench handles should not be extended by lengths of piping or other makeshift means.

16.1.23.   Wrenches should not be used on parts of machinery when moving.

16.1.24.   Shims should not be used with wrenches to make them fit.

16.1.25.   Wrenches should not be used as hammers unless specially constructed for this purpose.

16.1.26.   Files should be provided with well-fitting handles.

16.1.27.   Stakes or chisels being driven with a sledge hammer should be held by tongs and not by the hand.

16.2. Pneumatic tools

Construction

16.2.1.   Operating triggers on portable pneumatic tools should be:

(a) so placed as to minimise the risk of accidental starting of the machine; and

(b) so arranged as to close the air inlet valve automatically when the pressure of the operator's hand is removed.

16.2.2.   Hose and hose connections for compressed-air supply to portable pneumatic tools should be:

(a) designed for the pressure and service for which they are intended; and

(b) fastened securely to the pipe outlet and equipped with a safety chain.

16.2.3. Pneumatic shock tools should be equipped with safety clips or retainers to prevent dies and tools from being accidentally expelled from the barrel.

Use

16.2.4. Workers using portable power-driven tools should wear clothing conforming to the requirements of paragraph 13.3.2.

16.2.5. Tools should not be shot out of pneumatic hammers, but be removed by hand after use.

16.2.6. When cutting rivets with pneumatic cutters:

(*a*) the tools should be provided with a cage guard or other suitable device to catch the rivet heads; or

(*b*) the workers should be provided with suitable head and eye protection.

16.2.7. Pneumatic tools should be disconnected from the source of power and the pressure in hose lines released before any adjustments or repairs are made.

16.2.8. Before disconnecting hose lines, the air should be shut off.

16.2.9. Air supply lines should be adequately protected from damage by vehicles, etc.

16.2.10. Hoses should not be laid over ladders, steps, scaffolds, walkways, etc., so as to create a tripping hazard.

16.2.11. Compressed air should not be used for cleaning clothing or parts of the body.

16.3. Powder-actuated tools

Definitions

16.3.1. "Powder-actuated tool" means a device in which an explosive drives a projectile such as a nail or a stud into materials.

16.3.2. Powder-actuated tools are of three types:

(a) "high-velocity type" means a type of powder-actuated tool in which the projectile is driven directly by the gases from the explosive charge;

(b) "low-velocity piston type" means a type of powder-actuated tool in which the gases from the explosive charge drive a piston which propels the projectile;

(c) "hammer-operated low-velocity piston type" means a type of powder-actuated tool in which the piston is driven by a hammer blow in addition to the gases from the explosive charge.

General provisions

16.3.3.   The present section 16.3 applies to:

(a) high-velocity tools;

(b) low-velocity piston tools in which the projectile is struck while it is some way down the barrel of the tool; and

(c) subject to such modifications as the competent authority may accept, to hammer-operated low-velocity piston tools and other low-velocity tools.

16.3.4.   Whenever practicable, a low-velocity tool should be used instead of a high-velocity tool.

Tool construction

16.3.5.   Powder-actuated tools should have:

(a) a guard or protective shield that cannot be removed without rendering the tool inoperative;

(b) a device that prevents the tool from firing inadvertently, for example if it is dropped or while it is being loaded;

(c) a device that prevents the tool from firing if it is not approximately perpendicular to the working surface; and

(d) a device that prevents the tool from firing if the muzzle is not pressed against the working surface.

16.3.6.   Guards or shields should:

(a) be made of strong material; and

(b) be so designed as to arrest effectively ricocheting projectiles and fragments of projectiles and material.

16.3.7. Specially adapted guards or shields should be used when firing into corners, or into or at the side of projecting parts such as angle irons and wooden slats.

16.3.8. The recoil of a powder-actuated tool should not be capable of injuring the user even when the most powerful charge and the heaviest projectile are fired.

16.3.9. In normal conditions of use the noise of the detonation should not be such as to damage the ears in any way.

Cartridges

16.3.10. Only cartridges conforming to the maker's specifications should be fired in a powder-actuated tool.

16.3.11. The power of cartridges should be indicated on them, for instance by a distinctive colour.

Projectiles

16.3.12. Projectiles should be of a type and calibre exactly fitting the barrel of the tool.

16.3.13. Projectiles should be made of very hard and highly resilient metal.

16.3.14. The apex of the projectile should be so shaped as to offer the minimum resistance to penetration and the maximum resistance to extraction.

Inspection and maintenance

16.3.15. A powder-actuated tool should be inspected before each occasion of use to ensure that it is safe to use.

16.3.16. The inspection referred to in paragraph 16.3.15 should ensure in particular:

(a) that the safety devices are in proper working order;
(b) that the tool is clean;
(c) that all moving parts work easily; and
(d) that the barrel is unobstructed.

16.3.17. At intervals recommended by the manufacturer the tool should be completely dismantled and inspected for wear on the safety devices by a competent person.

16.3.18. Powder-actuated tools should only be repaired by the manufacturer or by competent persons.

16.3.19. Powder-actuated tools should be kept clean.

16.3.20. Tools found to be defective should be taken out of use.

Storage of tools, cartridges and projectiles

16.3.21. Powder-actuated tool cartridges should not be stored in an explosive atmosphere.

16.3.22. When not required for use, inspection or other purpose, powder-actuated tools should be kept in a suitable container.

16.3.23. Cartridges should be kept in a container that:

(a) is made of suitable material;

(b) is clearly marked to indicate its contents;

(c) is kept locked when not in use; and

(d) contains nothing except cartridges.

16.3.24. Cartridges of different strengths should not be kept together.

16.3.25. No tool should be stored loaded.

Use

16.3.26. Powder-actuated tools should be accompanied by instructions for their maintenance and use.

16.3.27. Only competent persons at least 18 years of age should use powder-actuated tools.

16.3.28. Operators of powder-actuated tools should wear safety goggles and, if necessary to prevent danger, wear safety helmets (hard hats) (when firing upwards), leggings (when firing downwards) or leather aprons (when firing forwards), ear protectors and be protected by face screens or shields.

16.3.29. Powder-actuated tools should not be loaded until they are to be used.

16.3.30. All powder-actuated tools should be treated as loaded until they have been examined to see whether they are loaded or not.

16.3.31. Powder-actuated tools should never be pointed at any person even when they are unloaded.

16.3.32. Powder-actuated tools should not be used in an explosive atmosphere.

16.3.33. Before a powder-actuated tool is fired:

(a) the user should make sure no one is in the danger zone; and
(b) if necessary to prevent danger, the zone should be protected by barricades, or danger notices should be affixed.

16.3.34. Powder-actuated tools should not be fired:

(a) into concrete or masonry near the edge;
(b) into existing holes unless a positive guide is used to ensure accurate alignment;
(c) into objects or structures through which the projectile could pass if that would cause danger;
(d) into resilient material that might cause projectiles to swerve or ricochet;
(e) into a projectile that is loose, deformed, broken or jammed; or
(f) in the vicinity of electrical or gas installations.

16.3.35. Powder-actuated tools should not be used on hard or brittle materials such as cast iron, hardened steel, glazed tiles, glass blocks or hard rock unless they are specially designed for such use.

16.3.36. When a powder-actuated tool is being fired:

(a) it should be held perpendicular to the working surface;
(b) it should have the muzzle pressed firmly against the working surface;
(c) it should if practicable be held with both hands; and
(d) the operator should have a firm and secure footing.

16.3.37. Cartridges used for a particular job should not be more powerful than the job requires.

16.3.38. Projectiles should be guided into the barrel to ensure perfect centring.

16.3.39. After firing, the tool should be examined and any foreign matter such as pieces of projectile or cartridge removed.

16.3.40. If a powder-actuated tool misfires:

(a) it should be pressed against the material for at least 15 seconds; and

(b) the cartridge should then be removed in strict accordance with the manufacturer's instructions.

16.3.41. Misfired cartridges should be kept in water until destroyed by a safe method.

16.3.42. Powder-actuated tools and cartridges should not be left unattended.

16.3.43. Powder-actuated tools should not be transported loaded or left loaded when not in use.

16.3.44. Cartridges should not be carried loose in clothing.

16.4. Electrical tools

16.4.1. Portable electrical tools should comply with the requirements of Chapter 17.

17. Electricity

17.1. Definitions

Protection against direct contact

17.1.1. "Protection against direct contact" means all measures designed to protect persons from hazards that arise when touching active parts of electrical equipment.

Protection against indirect contact

17.1.2. "Protection against indirect contact" means the protection of persons from hazards that may arise from contact with the normally dead parts of electrical equipment.

17.1.3. *Remarks:* A measure contributing to the protection against indirect contact is a means designed to ensure the fulfilment of this protective function (e.g. selection of appropriate arrangement or equipment).

Safety extra-low voltage

17.1.4. "Safety extra-low voltage" means a nominal voltage not exceeding 42 V between conductors, or, in the case of phase circuits, not exceeding 24 V between conductors and neutral, the no-load voltage of the circuit not exceeding 50 V and 29 V respectively.

17.1.5. *Remarks:* When safety extra-low voltage is obtained from supply mains of higher voltages it should be through a safety transformer or a convertor with separate windings.

17.2. General provisions

17.2.1. All parts of electrical installations should be of a standard of construction not lower from the safety point of view than national or international standard specifications approved or accepted by the competent authority.

17.2.2. All parts of electrical installations should be of adequate size and characteristics for the work they may be called upon to do and in particular they should:

(a) be of adequate mechanical strength to withstand working conditions in construction operations; and

(b) be not liable to damage by water, dust or electrical, thermal or chemical action to which they may be subjected in construction operations.

17.2.3. All parts of electrical installations should be so constructed, installed and maintained as to prevent danger of fire and external explosion.

17.2.4. All parts of electrical installations should be so constructed, installed and maintained as to prevent the danger of electric shock.

17.2.5. Personal protective equipment such as rubber gloves and rubber boots should not be considered as providing adequate protection against the risk of electric shock.

Identification

17.2.6. All electrical appliances and outlets should be clearly marked to indicate their purpose and voltage.

17.2.7. When the layout of an installation cannot be clearly distinguished, the circuits and appliances should be identified by labels or other effective means.

17.2.8. Circuits and appliances carrying different voltages in the same installation should be clearly distinguished by conspicuous means such as coloured markings.

Protection against excess voltage

17.2.9. Adequate precautions should be taken to prevent installations from receiving current at a higher voltage from other installations.

Protection against lightning

17.2.10. Where necessary to prevent danger, installations should be protected against lightning.

Signalling and telecommunication systems

17.2.11. Lines of signalling and telecommunication systems should not be laid on the same supports as medium and high voltage lines.

Protection against direct and indirect contact

17.2.12. Protection against excessive contact voltage in case of direct or indirect contact should be provided in all kinds of installations.

17.2.13. Taking into account the special requirements for any given circumstances (e.g. damp or wet places, working in tubes or tanks), the protection referred to in paragraph 17.2.12 may be afforded by one or more of the following:

(a) enclosure;
(b) complete insulation (double insulation, reinforced insulation);
(c) extra-low voltage;
(d) safety isolation (safety isolating transformer);
(e) earthing of the neutral;
(f) isolated neutral;
(g) earthing of the normally dead parts;
(h) current-operated earth-leakage circuit-breakers (high sensitive type);
(i) insulation control device.

17.2.14. No bare conductors or other bare current-carrying parts of equipment should be permitted unless adequate precautions are taken to prevent direct or indirect contact, for example by fencing or screening.

Enclosures, covers

17.2.15. Covers, protective mesh and housings should be made of incombustible material, possess adequate mechanical strength and be reliably secured.

17.2.16. The size of the openings of protective mesh or wire netting should be determined in relation to the distance to the nearest live parts.

17.2.17. It should not be possible to remove the enclosures of current-carrying parts without the use of a special tool unless the enclosure is interlocked.

Disconnection devices

17.2.18. A device that cuts off the current from all active conductors should be provided for every construction site.

17.2.19. A readily accessible device for cutting off the current from all active conductors should be provided in all circuits supplying consuming appliances.

17.2.20. As far as practicable, installations should be controlled by an appliance that automatically cuts off the current when an insulation defect occurs.

Lamps

17.2.21. Lamps and lamp fittings for general lighting should be installed not less than 2.5 m (8 ft) above the ground or floor wherever practicable. Lamps and lamp fittings which are within reach should be protected by a strong cover glass.

17.2.22. Outdoor fittings should be waterproof.

Flameproof equipment

17.2.23. Only flameproof equipment and conductors should be installed:

(a) at storeplaces for explosives or flammable liquids; and
(b) in explosive atmospheres.

Tunnels

17.2.24. Electrical installations in tunnels should comply with the relevant requirements of this chapter and the requirements of sections 32.8 and 32.9.

Welding

17.2.25. Electrical welding installations should comply with the relevant requirements of this chapter and the requirements of section 28.2.

Notices

17.2.26. A notice or notices should be kept exhibited at suitable places:

(a) prohibiting unauthorised persons from entering electrical equipment rooms;

(b) prohibiting unauthorised persons from handling or interfering with electrical apparatus;

(c) containing directions as to procedure in case of fire;

(d) containing directions as to the rescue of persons in contact with live conductors and the restoration of persons suffering from electric shock; and

(e) specifying the person to be notified in case of electrical accident or dangerous occurrence, and indicating how to communicate with him.

17.2.27. At all places where contact with or proximity to electric equipment can cause danger, suitable warnings should be placed.

Operation

17.2.28. Persons having to operate electrical equipment should be fully instructed as to any possible dangers of the equipment concerned.

17.3. Conductors

General provisions

17.3.1. All wiring should be supported on proper insulators, and not looped over nails, brackets, etc.

17.3.2. Neutral, compensating and protective conductors should be clearly distinguishable from other conductors.

17.3.3. Overhead lines should be carried on supports of adequate strength and at a height that prevents contact with persons, animals or equipment passing underneath.

17.3.4. Poles carrying electrical conductors or equipment should be securely anchored in the ground or to any other base.

17.3.5. While conductors are being removed from poles, the poles should be adequately guyed, so as to oppose one-sided pulls.

17.3.6. Outdoor conductors should be disconnectable by means of switches, fuses or plug and socket connections.

17.3.7. Overhead power lines carrying 440 V or more should have an adequate vertical clearance at places where they cross roads and other traffic areas.

17.3.8. Power lines in dangerous proximity to blasting operations should be de-energised during the blasting.

17.3.9. Conductors installed less than 2.5 m (8 ft) above the ground or floor should be fenced or enclosed in conduits of steel or other resistant material in order to be protected against damage.

17.3.10. As far as practicable, temporary conductors should not cross power lines, telephone lines or radio antennae.

17.3.11. Only conductors built to withstand rough treatment (heavy duty conductors) should be laid on the ground and, if necessary, they should be protected against damage from vehicles, mechanical equipment, rough handling, etc.

17.3.12. High-voltage insulated cables, such as those for power shovels and drag lines, should not be handled with the bare hands, but with rubber gloves or insulated appliances.

Third-rail systems

17.3.13. Third rails for travelling cranes, power shovels, etc. should be disconnectable.

17.3.14. If a number of machines are supplied from the same rail, each machine should be disconnectable on all poles.

17.3.15. If haulage machines are used with electric power shovels supplied from a third rail, the third rail should be outside the clearance gauge of the haulage machine.

17.3.16. Third-rail systems should be adequately protected so that machine operators and other persons cannot come into contact with the live rail.

Cable-reel haulage

17.3.17. Cables for cable-reel haulage should be equipped with a limit switch to prevent over-extension of the cable from the reel.

17.3.18. Reels should be protected by adequate enclosures or wire mesh fencing.

Flexible cables

17.3.19. If plug and socket connections are necessary for connecting cables to the mains, they should be:

(a) properly paired; and
(b) of adequate design.

17.3.20. Flexible cables for hand-held or portable apparatus should:

(a) contain an earthing conductor if the fed apparatus is protected by earthing;
(b) be protected against kinking by a steel spring, rubber tube or other suitable device at the motor end; and
(c) be relieved from mechanical strain at connections to terminals.

17.3.21. Hand-held apparatus and, where practicable, portable apparatus should be supplied by a single flexible cable.

17.3.22. All flexible cables should be maintained in good repair; they should not be joined except by means of appropriate plugs and sockets.

17.3.23. The flexible cable should not be used to lift a portable tool.

17.3.24. Flexible cables should not be left lying on surfaces that are oily or wet with corrosive liquids, unless they are covered with a suitably resistant material.

17.3.25. Flexible cables should be kept clear of loads, running gear and moving equipment.

17.3.26. Heavy rubber-insulated flexible cables should be used for extension lights for boilers, tanks and other places where they may be subjected to rough handling, or moisture.

17.4. Electrical equipment

General provisions

17.4.1. Control appliances such as switches, fuses and circuit-breakers should not be installed at places where there are explosives, flammable liquids or flammable gases unless they are flameproof.

17.4.2. Motors, distribution apparatus and switchgear should be protected against dripping and splashing water, particularly in pump rooms.

17.4.3. Unauthorised persons should not have access to electrical equipment rooms.

Transformers

17.4.4. Outdoor oil-filled transformers placed on the ground should be installed:

(a) at a place free from combustible materials; and
(b) so sunk below ground level or enclosed that escaping oil cannot spread.

17.4.5. As far as practicable, transformers on poles should be at least 4.5 m (15 ft) above the ground.

17.4.6. Transformers on poles less than 4.5 m (15 ft) above the ground should be adequately enclosed by fencing or other effective means.

Switchgear

17.4.7.   As far as practicable, switchgear should be enclosed by metal, plastic or other suitable material.

17.4.8.   If open-type outdoor switchgear has to be used:

(a) all live parts should be adequately protected against accidental contact by guards or by elevation;

(b) adequate working space should be provided around live parts; and

(c) the switchgear and associated transformers and other apparatus should be adequately enclosed.

17.4.9.   For control, instrument and protective relay circuits not associated with power circuits, dead front switchboards should be used.

Circuit-breakers

17.4.10.   Circuit-breakers should be of adequate breaking and making capacities to perform their normal function.

17.4.11.   Circuit-breakers should have their essential characteristics clearly marked on them.

17.4.12.   Except in extra-low voltage circuits as defined in section 17.1, the isolating device should act on all poles.

17.4.13.   It should not be possible for circuit-breakers to be opened or closed inadvertently by gravity or by mechanical impact.

Fuses

17.4.14.   Fuses should bear clear markings indicating their rated current, whether they are of the fast- or slow-breaking type and, as far as practicable, their rated breaking capacity.

17.4.15.   Effective protective measures should be taken to ensure that persons removing or inserting fuses will not be endangered, in particular by any adjacent live parts.

Switches

17.4.16.   All switches should be of the enclosed safety type.

17.4.17. Switches should be so installed and earthed as to prevent danger in their operation.

17.4.18. If switches can be closed by gravity they should be provided with a lock by which they can be kept open.

Motors

17.4.19. All motors should be equipped with a switch.

17.4.20. When a motor can be cut off from more than one place, where practicable, a stopping device should be provided in the immediate vicinity of the motor.

17.4.21. Motors should be so installed as to ensure that they can be adequately cooled.

17.4.22. Motors should be effectively protected against over-current.

Connections

17.4.23. At points where conductors are joined, branched or led into apparatus, they should be:

(a) mechanically protected; and
(b) properly and durably insulated.

17.4.24. Conductors should be joined, branched or led into apparatus through junction boxes, sleeves, bushings, glands or equivalent connecting devices.

17.4.25. Junction boxes or plug-and-socket couplings should be used for joining cables whenever practicable.

17.4.26. When parts of conductors are joined together, or conductors are joined to one another or to apparatus, the attachment should be made by screwing, clamping, soldering, riveting, brazing, crimping or equivalent means.

17.4.27. Junction boxes and connectors should be protected as far as possible against traffic, falls of ground, water and other sources of damage.

17.4.28. Whenever armoured cables are joined the junction boxes should be bridged by a suitable conductive bond between the armouring of the cables.

17.5. Transportable and portable electrical equipment

Hand-held and portable apparatus

17.5.1. The supply of electricity to hand-held apparatus should be at a voltage not exceeding 250 V.

17.5.2. Hand-held and portable machines should be equipped with a built-in switch.

17.5.3. Hand-held electrically operated tools should be provided with a built-in switch that will break the circuit automatically when the tool is released by the hands.

17.5.4. Portable electric tools, unless flame-proof, should not be used in flammable or explosive atmospheres.

Hand lamps

17.5.5. Hand lamps should be equipped with a strong cover of glass or other transparent material.

17.5.6. Portable lampholders should have:

(a) all current-carrying parts enclosed; and
(b) an insulated handle.

17.6. Electric trolley haulage

Locomotives

17.6.1. Current collectors should be so constructed that:

(a) they can be lowered safely and locked in the lowered position from the locomotive driver's cab; and
(b) live parts up to the collector shoe are protected against accidental contact.

17.6.2. There should be a disconnecting device between the current collector and the rest of the electrical equipment of the locomotive.

17.6.3. In tunnels, an adequate emergency lighting system should be provided in case of failure of the main current.

17.6.4. Drivers of trolley locomotives should be protected from contact with live conductors.

17.6.5. Controller handles should not be removable unless they are in the "off" position.

17.6.6. Controls of the dead man type should be provided where practicable.

17.6.7. Conductors and other electric parts on trolley locomotives should be protected against damage to their insulation from oil, heat or other causes.

17.6.8. Electrical brake circuits which do not use the motor current should:

(a) not include an automatic circuit-breaking device;
(b) be incapable of being made dead except by means of the controllers; and
(c) be incapable of being broken while the controller is in the "off" position.

17.6.9. Electrical braking devices should in all cases be supplemented by a powerful hand-brake which can be locked in position.

17.6.10. Locomotives should be provided with fire extinguishers of a type which is safe for use on live parts.

Trolley lines and tracks

17.6.11. Bare conductors for trolley lines and feeders should be so installed that they are protected as far as practicable against breakage.

17.6.12. At any point where persons must pass under trolley wires or bare feeders the lowest point of the wires or feeders should be at a safe height above the top of the rails, taking into account the height of persons and any article they may have to carry.

17.6.13. At any point where road vehicles must pass under trolley wires or bare feeders, the lowest point of the wires or feeders should be a safe height above the top of the rails, taking into account the size of vehicles.

17.6.14.   Trolley wires and feeders should be:

(a) fixed to suitably spaced insulating devices; and

(b) doubly insulated from the anchorage of the suspension device.

17.6.15.   Trolley wires and feeders should be protected collectively by an automatic circuit-breaker.

17.6.16.   Sectionalising switches should be placed at suitable intervals to permit trolley wires and feeders to be made dead as required.

17.6.17.   Tracks of trolley haulage systems should be effectively bonded at every joint and cross bonded at suitable intervals if they are used as return conductors.

17.7. Inspection, maintenance

17.7.1.   All electrical equipment should be inspected before it is taken into use to ensure that it is suitable for its proposed use.

17.7.2.   At the beginning of every shift every person using electrical equipment should make a careful external examination of the equipment and conductors for which he is responsible, especially flexible cables.

17.7.3.   Electrical conductors and equipment should only be repaired by electricians.

17.7.4.   As far as practicable no work should be done on live conductors or equipment.

17.7.5.   Before any work is begun on conductors or equipment that does not have to remain live:

(a) the current should be switched off;

(b) adequate precautions should be taken to prevent the current from being switched on again;

(c) the conductors or the equipment should be tested to ascertain that they are dead;

(d) the conductors and equipment should be earthed and short-circuited; and

(e) neighbouring live parts should be adequately protected against accidental contact.

17.7.6. After work has been done on conductors and equipment, the current should only be switched on again on the orders of a competent person.

17.7.7. Electricians should be supplied with sufficient adequate tools, and personal protective equipment such as rubber gloves, mats and blankets.

17.7.8. All conductors and equipment should be considered to be live unless there is certain proof of the contrary.

17.8. Work in the vicinity of electrical installations

17.8.1. When work is to be done in the neighbourhood of electrical conductors or installations, before it begins the employer should ascertain the voltages carried so that persons and equipment can be kept at a safe distance from the conductors or installations.

17.8.2. When any excavation is to be made or any borehole sunk, the employer should ascertain whether there are any underground conductors in, or in dangerous proximity to, the zone of operations.

17.8.3. No work should be done in dangerous proximity to a conductor or an installation until it has been made dead.

17.8.4. Before work begins, the electricity supply authority should certify that the conductor or installation has been made dead.

17.8.5. Before the current is restored, the employer should ensure that no worker remains on the work site.

17.8.6. If a conductor or an installation in the neighbourhood of which work is to be done cannot be made dead, special precautions should be taken and special instructions given to the workers so as to prevent danger.

17.8.7. As far as practicable, the precautions referred to in paragraph 17.8.6 should include adequately enclosing or fencing live conductors and installations.

17.8.8. If mobile equipment has to be employed in the neighbourhood of conductors or installations that cannot be made dead, its movements should be so controlled as to keep it at a safe distance from them.

18. Pressure plant

18.1. Steam boilers

18.1.1. Official regulations as regards the materials, design, construction, inspection and testing of steam boilers should be laid down.

18.1.2. Only competent persons should operate steam boilers.

18.1.3. Boiler feed water should be clean and free from foreign substances.

18.1.4. Steam should not be allowed to leak from the water column or its connections.

18.1.5. Boiler installations, especially gauges, should be well lighted.

18.1.6. The maximum working pressure should be marked in a distinctive colour on the pressure gauge.

18.1.7. The space around a boiler should be kept clear of obstructions and rubbish.

18.1.8. Boilers should preferably not be installed out of doors in cold countries.

18.1.9. Water should not be fed into a hot empty boiler.

18.1.10. If the water level is low the fire should be damped down, the ash door closed and the fire doors left open.

18.1.11. If persistent foaming occurs the boiler should be shut down.

18.1.12. Safety valves should operate freely at all times.

18.1.13. Water gauge glasses and water columns should be blown out at frequent intervals during each shift to make sure that all connections are clear.

18.1.14. Blow-off cocks should be opened and closed slowly so as to avoid water hammer.

18.1.15.   Boilers should be blown off into a sump or pit, or other effective precautions should be taken to avoid scalding persons.

18.1.16.   Scale should not be allowed to accumulate in boilers.

18.1.17.   High pressure steam boilers in operation should not be left unattended.

18.1.18.   All working parts of steam boilers such as valves, cocks, injectors and pumps should be frequently inspected by the operator.

18.1.19.   Boilers should only be repaired by competent persons after all pressure has been removed.

18.2. Compressors

Construction

18.2.1.   Compressors should be provided with a plate or the like indicating:

(a) the year of construction;
(b) the capacity per minute or hour;
(c) the pressure in atm. gauge;
(d) the number of revolutions per minute; and
(e) the power.

18.2.2.   Compressors should be equipped with:

(a) automatic devices that will prevent the maximum safe discharge pressure from being exceeded;
(b) a quick-release valve; and
(c) suitable arrangements for preventing contamination where persons are working in confined spaces.

18.2.3.   Stationary compressors should be installed on firm foundations and firmly fastened in place.

18.2.4.   All moving parts of compressors should be effectively protected against accidental contact.

18.2.5.   Compressors and their accessories should be protected against liquid hammer.

18.2.6. Pressure gauges for high pressures should be so protected that no person can be injured if they burst.

18.2.7. Compressors in which explosive mixtures of gas may form should be protected against sparking.

18.2.8. Where compressor cylinders are equipped with water-cooling jackets it should be possible to observe the water flow.

18.2.9. Intercoolers and aftercoolers should be able to withstand safely the maximum pressure in the air-discharge piping.

Air piping

18.2.10. Where necessary to prevent danger, air-discharge piping of compressors should be provided with:

(a) a fusible plug; and

(b) insulating covers to protect workers against burns, and to prevent fire risks.

18.2.11. Where necessary to prevent danger, an oil separator should be provided between the compressor and the air receiver.

18.2.12. Where stop valves are installed in air-discharge piping:

(a) they should be easily accessible for inspection and cleaning; and

(b) one or more safety valves should be installed between the compressor and the stop valve.

Steam and gas lines

18.2.13. Steam or gas lines to steam-driven or gas-driven compressors should be provided with a manually operable throttle valve in a readily accessible position.

18.2.14. The open and closed position of valves on steam and gas lines and air-discharge piping should be clearly marked.

Steam piping

18.2.15. Temporary steam piping should be:

(a) securely supported;

(b) adequately insulated, screened or guarded at places where workers could come into contact with it; and

(c) clearly marked.

Operation

18.2.16. Compressors should only be operated by persons fully instructed as to any possible dangers of the equipment.

18.2.17. On or near compressors instructions for their operation should be displayed.

18.2.18. Air supplied to compressors should be clean and free from any explosive, flammable or toxic contaminants.

18.2.19. All working parts, including speed governors, safety valves and oil separators should be inspected and, if necessary to prevent danger, cleaned at suitable intervals.

18.2.20. Only a cleaning agent specified by the manufacturer should be introduced into compressor cylinders and connected piping.

18.2.21. Leaky valves should be promptly repaired or replaced.

18.2.22. Pipes should not be connected or disconnected while there is any pressure in them unless specially equipped for this purpose.

18.2.23. Care should be taken that lubricating oil does not spread to coolers, receivers and other parts of the system where it might cause danger.

18.2.24. Dangerous gases escaping from safety and other valves should be led off safely.

18.3. Air receivers

18.3.1. Official regulations as regards the materials, design, construction, inspection and testing of air receivers should be laid down.

18.3.2. Air receivers should be equipped with:

(a) a safety valve;

(b) a pressure gauge; and

(c) a drain cock.

18.3.3.   Air receivers should be provided with suitable openings for inspection and cleaning.

18.3.4.   Air receivers should be so installed as to be:

(a) protected from the weather; and

(b) accessible for thorough inspection.

18.3.5.   Air receivers should be examined and tested at appropriate intervals by a competent person.

18.3.6.   The maximum safe pressure should be marked in a distinctive colour on the pressure gauge.

18.3.7.   Where necessary to prevent danger, a pressure-reducing valve or a stop valve, or both, should be inserted in the piping between the air receiver and the compressor.

18.3.8.   Between the receiver and each consuming appliance there should be a stop valve.

18.3.9.   Air receivers should be cleaned of oil, carbon and other foreign substances at suitable intervals.

18.4. Gas cylinders

General provisions

18.4.1.   Cylinders for compressed, dissolved or liquefied gases should be properly constructed with sound material.

18.4.2.   Cylinders for compressed, dissolved or liquefied gases should be distinctively identified as to their contents, to ensure the safety of the worker.

18.4.3.   No gas cylinder should be used unless it is fitted with:

(a) a high-pressure gauge;

(b) a reducing valve with pressure regulator and safety relief device; and

(c) a low-pressure gauge.

Inspection, testing

18.4.4.   Gas cylinders should be inspected and tested by a competent person:

(a) before being taken into use for the first time;
(b) before being taken into use after repairs; and
(c) at suitable intervals.

Storage

18.4.5.   Cylinders should be adequately protected against excessive variations of temperature, direct rays of the sun, accumulation of snow and continuous dampness.

18.4.6.   Store rooms containing charged cylinders should be conspicuously marked on the outside with suitable danger signs.

18.4.7.   When stationary, all cylinders (whether loaded or empty) should be maintained in an upright position.

18.4.8.   Store rooms should be adequately ventilated.

18.4.9.   No person should smoke in a cylinder store room.

18.4.10.   Cylinders should be segregated for storage by type of gas.

18.4.11.   If necessary to prevent danger, there should be a fire-resisting partition between cylinders of oxygen and cylinders of acetylene or fuel gas.

18.4.12.   Except when in use, cylinders containing combustible gases should not be kept in rooms where welding or cutting work is being done, and oxygen cylinders should be kept separated from all other cylinders.

18.4.13.   Empty cylinders should be kept apart from charged cylinders.

18.4.14.   Leaky cylinders charged with acetylene or liquefied fuel gas should be taken into the open air at a safe distance from any open flame or sparks.

18.4.15. Cylinders should be kept at a safe distance from:

(a) electrical conductors such as third rails, trolley wires and lightning conductors; and

(b) all operations which produce flames, sparks or molten metal or cause excessive heating of the cylinders.

18.4.16. When gas cylinders are stored inside buildings:

(a) the number of cylinders should be as small as practicable;

(b) the cylinders should be stored in rooms with fire-resisting walls;

(c) the cylinders should be kept at a safe distance from flammable substances, radiators and other sources of heat; and

(d) the cylinders should be secured against falling and rolling.

18.4.17. Cylinders of heavy petroleum gases should not be stored underground except when it is absolutely unavoidable.

18.4.18. Valve protection caps should always be in place when cylinders are not in use or not connected for use.

18.4.19. No tools or other objects should be placed on the top of a gas cylinder.

Handling

18.4.20. Cylinders should not be knocked, dropped or rolled in handling, or otherwise subjected to violent shocks.

18.4.21. The valves of cylinders should not be opened by hammering or other violent means and should always be opened slowly.

18.4.22. Acetylene cylinders should be opened slowly with a special tool which should be left on the stem so that the valve can be closed quickly in an emergency.

18.4.23. If cylinders charged with liquefied gases are heated for emptying, this should be done with a water jacket and not with an open flame.

18.4.24. The valves of cylinders should be closed immediately after emptying.

18.4.25. Oxygen cylinders should not be allowed to come into contact with oil or grease.

18.4.26. Oxygen under pressure should not be allowed to come into contact with oily or greasy surfaces such as clothes or containers.

Transport

18.4.27. Suitably designed equipment should be used for transporting gas cylinders on construction sites.

18.4.28. When cylinders are moved by a hoisting mechanism, a properly designed cradle or the like should be used.

18.4.29. Cylinders should not be hoisted by slings, hooks or magnets.

Use of gas

18.4.30. Gas welding and cutting should be done in accordance with the requirements of section 28.1.

18.5. Acetylene generators

18.5.1. Official regulations should be laid down for acetylene generators.[1]

---

[1] When no regulations exist, acetylene generators should comply with the relevant requirements of the *Model Code of Safety Regulations for Industrial Establishments* issued by the International Labour Office.

19. Floating operational equipment

19.1. General provisions

19.1.1. The following provisions apply to all floating operational equipment and all land equipment which is used on water.

19.1.2. Gangways, pontoons, bridges, footbridges and other walkways or workplaces over water should:

(a) possess adequate strength, stability and buoyancy;
(b) be sufficiently wide to allow safe movement of workers;
(c) have level surfaces free from protruding knots, bark, nails, bolts and other tripping hazards;
(d) if necessary to prevent danger, be boarded over;
(e) if necessary to prevent danger, be adequately lighted when natural lighting is insufficient;
(f) be provided at appropriate points with sufficient lifebuoys, lifelines and other life-saving equipment;
(g) where practicable and necessary to prevent danger, be provided with toe-boards, guard-rails, hand ropes or the like;
(h) be kept clear of tackle, tools and other obstructions; and
(i) be strewn with sand, ashes or the like when made slippery by ice.

19.1.3. Floating structures should, if necessary to ensure protection, be provided with shelters.

19.1.4. Floating structures accommodating machinery should be provided with fencing or guard-rails and toe-boards on all sides.

19.1.5. Floating operational equipment should be provided with sufficient and suitable rescue equipment such as lifelines, gaffs and ring buoys.

19.1.6. Ring buoys should always be kept readily accessible on the deck of floating operational equipment.

19.1.7.   Where rafts are used, they should:

(a) be strong enough to support safely the maximum loads that they will have to carry;

(b) be securely moored; and

(c) have safe means of access.

19.1.8.   Decks of floating equipment should as far as practicable be enclosed by guard-rails and toe-boards complying with the requirements in paragraphs 2.6.1 to 2.6.5.

19.1.9.   Iron decks should be studded or have some other type of non-slip surface.

19.1.10.   The steersman of floating operational equipment should have an unobstructed view.

19.1.11.   Floating operational equipment, including boats, should not be overloaded.

19.1.12.   Unauthorised persons should not be allowed on any floating structure.

19.1.13.   Ice should not be allowed to accumulate on any floating operational equipment.

19.1.14.   Slippery surfaces should be made safe where necessary.

19.1.15.   As far as practicable, deck openings for buckets should be fenced.

19.1.16.   For crossing deck openings, gangways complying with the requirements of section 3.3 should be provided.

19.1.17.   A safe walkway should be provided on all floating pipelines.

19.1.18.   No person should enter a hydraulic dredge gear room without first informing the leverman and without being accompanied by a second person.

19.1.19.   Hoist lines, drag lines, buckets, cutter heads and bridles should be inspected daily.

19.1.20.   Before dredging operations begin, the position of underwater equipment such as cables and pipelines should be ascertained.

19.1.21. Where anchors are used, adequate means should be provided to raise or lower them.

19.1.22. Workers should be embarked and disembarked only at suitable and safe landing places.

19.2. Boats

19.2.1. Boats used to transport workers by water should comply with requirements which should be laid down by the competent authority.

19.2.2. Boats used to transport workers should be manned by an adequate and experienced crew.

19.2.3. The maximum number of persons transported in a boat should not be greater than safety allows and this number should be displayed in a conspicuous place.

19.2.4. On boats, suitable and adequate life-saving appliances should be provided and properly placed and maintained.

19.2.5. Tow-boats should have a device by which the tow-rope can be quickly released.

19.2.6. Power-driven boats should carry suitable fire extinguishers.

19.2.7. Row-boats should carry a spare set of oars.

20. Silos

20.1. Construction and equipment of silos

General provisions

20.1.1. Silos should:

(a) be erected on adequate foundations; and
(b) withstand the stresses to which they will be subjected without any deformation of walls, floors or other load-bearing parts.

20.1.2. Lanes for trackless vehicles that drive under the discharge openings or appliances should allow an adequate clearance between the vehicles and the silo.

20.1.3. Traffic lanes under silos should be clearly and durably marked with the maximum width and height of vehicles allowed to use them.

20.1.4. All places in silos to which workers have to go should be provided with safe means of access such as stairs, fixed ladders, columns of iron rungs, gangways, boatswain's chairs or hoists complying with the relevant requirements of this Code.

20.1.5. Inside silos there should be no structure or projection that hinders the flow of material.

20.1.6. Facilities should be provided to enable the quantity of material in the silo to be assessed without entering the silo.

20.1.7. In closed parts of silos, only heating appliances with no open flames should be used.

20.1.8. If silos have electric lighting the switches and sockets should be outside near the entrances.

20.1.9. On silos, notices should be conspicuously displayed:

(a) containing details of the requirements for entry contained in paragraph 20.2.2; and
(b) calling attention to the danger of sinking in fine materials.

Filling arrangements

20.1.10. Filling openings on the tops of silos should be so protected by grids or other effective means that no person can fall in.

20.1.11. If the silo is filled by grabs, the top gangways and platforms should be 1 m (3 ft 3 in) below the edge.

Blockages

20.1.12. If the material in the silo is liable to cause a blockage, agitators, compressed air or other mechanical devices should be preferably provided for clearing it. Equipment such as poles, long-handled tools, rammers or scraper chains should also be available for emergency use.

20.1.13. If equipment for clearing the silo has to be used from the top, gangways or platforms complying with the requirements of sections 3.2 or 3.3 should be provided.

20.1.14. Entrances in the side walls for clearing blockages and located higher than 1.5 m (5 ft) above the bottom should be provided with a platform complying with the requirements of section 3.2.

Dangerous contents

20.1.15. All silos should be adequately ventilated; the ventilation should take into account any gases which may be produced.

20.1.16. Ventilation openings should be protected against freezing, rain and snow.

20.1.17. Silos for slaked lime should be equipped with a pressure release device.

20.1.18. Silos for material liable to spontaneous combustion should be provided with fire-extinguishing equipment.

20.1.19. Silos for harmful dusty material should:

(a) be dust-tight;

(b) have air-tight filling arrangements; and

(c) be provided with dust-exhaust equipment at discharge points.

20.1.20. When workers are required to work at bag-dumping positions:

(a) suitable respirators should be provided and used; and

(b) the pit or other opening should be covered by a grating.

20.1.21. In silos in which explosive mixtures of gases or dusts are liable to form:

(a) all electrical equipment including hand lamps should be flameproof;

(b) only non-sparking tools should be used; and

(c) explosion vents should be provided in the walls.

Discharge arrangements

20.1.22. If the discharge is automatic, the sloping floor and the sides of the hopper should be steeper than the angle of repose of the material.

20.1.23. If the silo has a level floor there should be a sufficient number of discharge openings or appliances.

20.1.24. Discharge openings in the side walls of silos should be provided on the inside with a ledge that prevents dangerous quantities of the material from falling on to workers employed at the opening.

20.1.25. Controls of gates closing discharge openings should:

(a) be easy to operate;

(b) be installed in a safe place from which the discharge can be watched; and

(c) if necessary to prevent danger, be capable of locking the gates closed.

20.2. Operation of silos

20.2.1. Entrances of silos should be kept closed and locked.

20.2.2. No worker should enter a silo unless:

(a) the discharge opening is closed and secured against opening and filling is stopped;

(b) he is duly authorised to do so;

(c) he is under constant surveillance by another authorised person; and

(d) he wears a safety belt with a lifeline securely attached to a fixed object and complying with the requirements of Chapter 36.

20.2.3. Rope ladders should not be used in silos.

20.2.4. Blockages in silos should be cleared from above rather than from the discharge openings.

20.2.5. If silos have to be entered to clear blockages:

(a) only competent persons should be employed;

(b) no material should be put into or taken out of the silo while any person is inside; and

(c) if necessary to prevent danger, persons entering the silo should be equipped with a suitable respirator.

20.2.6. Silos should not be entered through the discharge openings unless the silos are quite empty.

20.2.7. If repair work has to be done on silos that are not quite empty, working platforms complying with the requirements of section 3.2 should be provided or other suitable precautions should be taken to provide a safe working place.

21. Dangerous substances and radiations

21.1. General provisions

21.1.1. Harmful atmospheric contaminants such as dusts, fibres, fumes, gases and mists should be rendered harmless as near their point of origin as practicable by removal, suppression or other effective means.

21.1.2. When harmful atmospheric contaminants cannot be rendered harmless, workers exposed to them should be provided with respiratory protective equipment complying with the relevant requirements of paragraphs 36.1.38 to 36.1.46.

21.1.3. Where necessary to prevent danger, the atmosphere of workplaces should be tested for harmful contaminants at suitable intervals by a competent person.

Confined spaces

21.1.4. No person should enter a confined space such as a room, pit, shaft, sewer or tank in which it is reasonable to expect that toxic, asphyxiating, flammable or other dangerous gases may be present or may accumulate, or oxygen may be deficient, unless:

(a) the atmosphere has been found to be safe after a suitable test and such tests are repeated at suitable intervals; and

(b) adequate ventilation is provided.

21.1.5. If the conditions in the preceding paragraph cannot conveniently be fulfilled, persons may enter such spaces for prescribed periods using a hose mask or self-contained breathing device.

21.1.6. While a worker is in a confined space:

(a) adequate facilities should be readily available for rescuing him;

(b) an attendant or attendants should be stationed at or near the opening; and

(c) suitable means of communication should be maintained between the worker and the attendant or attendants.

Air-cleaning equipment

21.1.7.   Air-cleaning equipment should be so placed that:

(a) collected contaminants can be removed from it without causing danger; and

(b) it can be serviced and repaired without causing danger by recontamination of the atmosphere.

21.1.8.   Atmospheric contaminants removed by exhaust systems should not be discharged so that they can recontaminate the atmosphere of workplaces.

Containers

21.1.9.   Containers in which there are dangerous substances should be:

(a) plainly marked to indicate the contents and their dangerous nature; and

(b) carry or be accompanied by instructions for safely handling the contents.

21.2. Highly combustible materials

General provisions

21.2.1.   Highly combustible solids and flammable liquids should only be stored in locked storeplaces that are not occupied by persons.

21.2.2.   Fire-protection measures should be taken at storeplaces for combustible and flammable materials in conformity with the requirements of section 2.4.

Flammable liquids

21.2.3.   Buildings or structures for the storage of flammable liquids should be adequately vented.

21.2.4.   Buildings or structures for the storage of flammable liquids in bulk should:

(a) be surrounded by a watertight wall or bank, or by a pit, such that all the liquid stored could be retained if it escaped; or

(b) be so constructed that no liquid could escape as the result of a fire or other occurrence.

21.2.5.   When not stored in bulk, flammable liquids should be kept in containers that:

(a) are tightly closed;

(b) are fireproof and unbreakable; and

(c) are labelled to indicate their contents.

21.2.6.   When flammable liquids are transferred from one bulk container to another, the two containers should be electrically bonded and earthed so as to prevent danger from static electricity.

21.2.7.   During long periods of disuse, or before storage, all containers that have held flammable liquids should be rendered free from any residual flammable contents.

21.2.8.   Proper precautions should be taken, including the thorough cleansing of flammable materials from containers, before any heat is applied, and any repairs should be carried out in the open and outside any buildings.

21.2.9.   Petrol (gasoline) should not be used for removing grease or other substances from equipment, materials or the body.

Gases, vapours, dusts

21.2.10.   In enclosed spaces in which flammable gases, vapours or dusts can cause danger:

(a) there should be no lighting equipment, including hand torches, that is not flame-proof;

(b) there should be no open fire or flame;

(c) adequate ventilation should be provided; and

(d) no person should smoke.

Explosives

21.2.11.   Commercial explosives should be handled in compliance with the requirements of Chapter 22.

Bituminous substances

21.2.12.   When hot asphalt, tar or other bituminous materials are used, the requirements of section 15.5 and section 29.7 should be complied with.

Laying floors, etc.

21.2.13.   When flammable substances are used for laying floors, facing walls, etc., the requirements of section 29.9 should be complied with.

21.3. Toxic and irritant substances

21.3.1.   Workers exposed to toxic or irritant substances should be provided with personal protective equipment (including clothing and respirators) if necessary to prevent danger.

21.3.2.   Working protective equipment should be:

(a) cleaned, and if necessary to prevent danger, sterilised at suitable intervals; and

(b) maintained in good repair.

21.3.3.   If necessary to prevent danger, working clothing and coverings should not be removed from the work site by the users.

21.3.4.   Where necessary, personal protective equipment and coverings should be removed before the wearer partakes of any food.

21.3.5.   All workers exposed to toxic substances should thoroughly wash the hands and face before partaking of any food or, where necessary, before leaving the work site.

21.3.6.   Workers exposed to toxic or irritant substances should promptly report any physical complaints to the medical service, first-aid post or a supervisor.

21.3.7.   Persons with open wounds should not handle toxic or corrosive substances.

21.3.8.   For drawing off acids from containers, siphons, tipping appliances or other suitable devices should be used.

21.3.9.   Approaches to working areas where danger from toxic or irritant substances may exist should be provided with notices or signs:

(a) naming the substance involved; and

(b) indicating the type of respiratory equipment to be worn.

21.3.10.   When wood preservatives are used the requirements of section 29.8 should be complied with.

21.3.11.   When materials containing asbestos, glass wool, glass fibres, mineral wool, silica dust or the like are used, the requirements of section 29.10 should be complied with.

21.4. Ionising radiations and laser radiations

21.4.1.   Official regulations should be laid down for operation on the site, storage, handling and use of radioactive substances or apparatus generating ionising radiations or laser radiations.

21.4.2.   Where official regulations do not exist, radioactive materials and devices should only be handled, used or stored under the guidance of an expert.

22. Transport, storage and handling of explosives

22.1. General provisions

22.1.1. Only competent persons specially authorised for the purpose should handle or use explosives.

22.1.2. Deteriorated or greasy explosives should not be used.

22.1.3. All explosives issued from a magazine should be accounted for, and unused explosives should be returned to the same magazine on the completion of the operation for which they were drawn.

22.1.4. Workers storing, transporting or otherwise handling explosives should not smoke or carry open lights.

22.2. Transport of explosives

General provisions

22.2.1. Road and rail vehicles used to transport explosives should:

(a) be in good condition and running order;

(b) have a tight wooden or non-sparking metal floor;

(c) have sides and ends high enough to prevent the explosives from falling out;

(d) in the case of road vehicles, carry at least two suitable fire extinguishers; and

(e) be plainly marked by a red flag, lettering or otherwise to indicate that they are carrying explosives.

22.2.2. Explosives being transported in vehicles should:

(a) not be transported together with metal, flammable or corrosive substances;

(b) not be allowed to come into contact with sparking metal; and
(c) be effectively separated from any detonators carried in the same vehicle.

22.2.3.   Persons travelling on vehicles transporting explosives should not smoke or have open flames or lights.

22.2.4.   Unauthorised persons should not travel on vehicles transporting explosives.

Containers

22.2.5.   Explosives and detonators should be transported from the magazine to the workplace in their original containers separately or in special closed containers of non-sparking material.

22.2.6.   Different types of explosive should not be transported in the same container.

22.2.7.   Containers should be marked to show the type of explosive kept in them.

Trains

22.2.8.   When a train transports explosives, only workers in charge of the transport should travel in it.

22.2.9.   When a train transports both explosives and detonators, they should be placed in different vehicles or be separated by equivalent means.

22.2.10.   No worker should travel in a vehicle transporting detonators.

Road vehicles

22.2.11.   Road vehicles without springs should not be used for transporting explosives.

22.2.12.   Road vehicles transporting explosives should not be left unattended.

22.2.13.   Explosives should not be transported in trailers; when explosives are transported in semi-trailer trucks the trailer should be equipped with safety chains.

22.2.14. Road vehicles carrying explosives should come to a full stop before crossing unprotected railway tracks or entering or crossing intersecting roads.

22.2.15. When refuelling road vehicles carrying explosives, adequate precautions should be taken against fire.

Boats

22.2.16. Boats used to transport explosives should be plainly marked by a red flag, lettering or otherwise, to indicate that they are carrying explosives.

22.3. Storage of explosives

General provisions

22.3.1. Explosives should be permanently stored only in magazines which should:

(a) be at a safe distance from occupied buildings or areas;

(b) be substantially constructed, bullet-proof and fire-resistant;

(c) be clean, dry, well ventilated and cool; and

(d) be kept securely locked.

22.3.2. Only flame-proof electric lighting equipment should be allowed in explosives magazines.

22.3.3. Explosives should not be stored together with detonators or any primed explosives.

22.3.4. No flammable substances or sparking metal objects should be stored or used in explosives magazines.

22.3.5. In explosives magazines or in a restricted and clearly marked zone around them:

(a) there should be no smoking, matches, open lights or open flames;

(b) no firearms should be discharged; and

(c) no flammable debris such as grass, leaves or brushwood should be allowed to accumulate.

22.3.6. If nitroglycerin has leaked from deteriorated explosives in a magazine, the floor should be thoroughly washed with an agent approved by the manufacturer.

22.3.7. Only persons authorised to handle explosives should have the keys of magazines.

22.3.8. Explosives magazines should not be opened during or on the approach of an electrical storm.

22.3.9. Where necessary to protect explosives against white ants and vermin:

(a) boxes of explosives should be stacked on trestles;
(b) wooden trestles should stand in termite-resistant containers filled with water; and
(c) openings in magazines, such as ventilators, should be provided with protection against the entrance of squirrels, rodents, snakes, etc.

Temporary storage

22.3.10. If quantities of explosives and detonators have to be provisionally stored outside the main magazine special accommodation should be provided, such as a special room, a portable magazine or a suitable container.

22.3.11. Temporary store rooms for explosives should have no fireplace or flue.

22.3.12. In temporary store rooms for explosives:

(a) nothing but explosives or detonators should be stored; and
(b) explosives should be kept separate from detonators.

22.3.13. Only one type of explosive should be kept in a temporary container.

22.3.14. Temporary magazines of explosives should be at a safe distance from living and work rooms and blasting operations.

22.3.15. Temporary containers for powder explosives should have no metal parts inside.

22.3.16. Temporary store rooms and cases for explosives should be kept locked when not in use.

22.3.17. Only persons authorised to handle explosives should have the keys of temporary store rooms or cases for explosives.

22.4. Handling of explosives

22.4.1. Containers of explosives should not be opened with sparking tools, provided that metal slitters may be used to open cartons or similar containers.

22.4.2. Explosives should be kept away from open flames, sparks and excessive heat.

22.4.3. Explosives should be protected from impact.

22.4.4. Explosives and detonators should be kept in their containers when not in use.

22.4.5. Containers of explosives should be kept closed when not in use.

22.4.6. Explosives should not be carried in pockets or elsewhere on the person.

22.4.7. Cartridges and primers should not be made up in magazines or near large quantities of explosives.

22.4.8. Detonators should not be treated roughly or tampered with.

22.4.9. As soon as the approach of an electrical storm is detected, all workers should be removed from the area where explosives are stored or are in use.

22.4.10. Frozen explosives should not be broken, rubbed, cut, squeezed, struck or otherwise roughly treated.

22.4.11. The thawing of frozen explosives should only be carried out under the guidance of a specialist in accordance with the following requirements:

(a) frozen explosives should be thawed in a water jacket containing warm water, not hotter than 50 °C;

(b) the explosive should not come into contact with the water;

(c) thawing should be carried out at a safe distance from fireplaces, stoves, steam pipes, boilers and other sources of heat; and *(d)* all workers not employed on thawing should keep at a safe distance.

22.4.12.   Frozen primed cartridges should not be thawed but destroyed.

22.4.13.   If black-powder cartridges are made up on the construction site this should be done in an isolated place away from all magazines, at a safe distance from sources of heat or sparking, with non-sparking appliances and in daylight.

22.4.14.   Black powder should be prevented from spreading over the ground or on clothes.

22.5. Disposal of explosives

22.5.1.   No explosives should be left lying about without supervision.

22.5.2.   Explosives should not be destroyed except in conformity with the manufacturers' instructions.

22.5.3.   No material used in the wrapping or packing of explosives should be burned in a stove, fireplace or other confined space.

22.5.4.   No person should remain within 30 m (100 ft) of a fire in which wrapping or packing material is burned.

23. Blasting

23.1. General provisions

23.1.1. Blasting caps, safety fuses, wiring and other blasting equipment should conform to specifications to be laid down in national or other official regulations or standards.

23.1.2. Dynamite should not be removed from its original wrapper until it is being loaded into boreholes.

23.1.3. Detonating fuses, electric blasting caps or caps properly crimped to safety fuse may be used in wet boreholes.

23.1.4. Detonating fuses or electric blasting caps should be used in underwater blasting.

23.1.5. Underwater blasting operations should comply with the requirements of paragraphs 34.4.17 to 34.4.30.

23.1.6. As far as practicable, blasting should be done off shift or during breaks in the work.

23.1.7. As far as practicable, blasting above ground should be done in daylight.

23.1.8. If blasting above ground has to be done in the dark, roadways and pathways should be adequately lighted by artificial light.

23.1.9. If blasting can endanger workers in another undertaking:

(a) blasting times should be agreed between the two undertakings; and

(b) no shot should be fired until a warning has been given to the other undertaking and acknowledged by it.

23.1.10. Loaded boreholes should not be left unattended after the end of the shift.

23.1.11. Arrangements should be made for advanced warning to be given of the approach of any electrical storms.

23.1.12. When an electrical storm is approaching, all work with explosives should be stopped and all workers should be moved from any area where explosives are in use.

23.1.13. Suitable and sufficient means of egress to ground level should be provided in all shafts, boreholes, excavations, trenches, and all other places where explosives are handled above or below ground level.

23.1.14. At an appropriate time before the final blasting warning, workers in the area should be removed to a designated safe place.

23.1.15. An unmistakable, audible, final warning should be sounded one minute prior to the detonation of explosives; after completion, when the person in charge has established that safe conditions prevail, an "all clear" should be sounded.

23.1.16. To prevent persons entering any danger zone during blasting operations:

(a) look-outs should be posted around the area of operations;
(b) warning flags should be flown; and
(c) conspicuous notices should be posted at points around the area of operations.

23.1.17. Notices referred to in paragraph 23.1.16 should indicate:

(a) that explosives are in use;
(b) the audible warning sound and the "all clear", and state when they will be sounded; and
(c) the warning flags in use, including an "all clear" flag.

23.2. Drilling and loading boreholes

23.2.1. Before a borehole is loaded all workers not employed in the blasting operation should withdraw to a safe place.

23.2.2. No smoking or open flame should be allowed in the loading area.

23.2.3. In underground blasting, the charge of explosive should consist of:

(a) a single cartridge, or
(b) a row of cartridges all in contact with one another or with a detonating fuse.

23.2.4. In underground blasting:

(a) charges should only be fired with one primed cartridge and one detonator; and
(b) the detonator should be at one of the ends of the charge.

23.2.5. Blasting caps should be crimped by an appliance authorised for use in mines or otherwise approved by the competent authority.

23.2.6. Blasting caps should not be crimped by biting.

23.2.7. When black-powder cartridges are fired with fuse the cartridge attached to the fuse should be the last to be loaded.

23.2.8. Boreholes should be checked with a wooden tamping bar or similar device to ensure that they are safe for the insertion of the charge.

23.2.9. When boreholes have been drilled or sprung they should not be loaded unless:

(a) they have cooled sufficiently;
(b) they are free of hot metal, or burning or smouldering material; and
(c) they have been cleaned by being blown out with compressed air or by other effective means.

23.2.10. Blow pipes for cleaning out holes should be of non-sparking material.

23.2.11. Boreholes should be large enough throughout their length to enable cartridges to be inserted easily.

23.2.12. Boreholes should not be drilled into other boreholes, misfires or sockets.

23.2.13. Boreholes should not be sprung near other boreholes if they are loaded.

23.2.14.   No borehole that has previously contained explosives should be deepened.

23.2.15.   When blasting in the presence of water or acid, a plastic-type safety fuse or, if available, a tape-covered fuse should be used.

23.2.16.   Explosives should not be forced into boreholes.

23.2.17.   Detonators should not be forced into dynamite.

23.2.18.   Primers should not be handled roughly, slit or deformed.

23.2.19.   Loose powder should be loaded into the borehole with a funnel, provided that the funnel is made of non-sparking material.

23.2.20.   As soon as a borehole is loaded, the blaster should remove surplus explosive and detonators to the store place.

23.2.21.   With delayed-action detonators the primed cartridge should be placed at the bottom of the borehole.

23.2.22.   Except in the case of a misfire a loaded hole should not be untamped or unloaded.

23.2.23.   When sticks or other devices are used to load a borehole they should be of wood or other suitable non-sparking material.

23.3. Tamping

23.3.1.   Tamping material should not contain any hard ingredients such as stones.

23.3.2.   Tamping bars should be of wood or other suitable non-sparking material.

23.3.3.   Charges should be tamped gently.

23.3.4.   Primers should not be rammed.

23.3.5.   Boreholes should be tamped with an adequate thickness of sand, earth, clay or other suitable incombustible material.

23.3.6.   Care should be taken not to kink or damage fuses or detonator wires when tamping.

23.4. Firing: general provisions

23.4.1. Shots should not be fired unless:

(a) adequate warning has been given to all persons who might be endangered and they have taken shelter;

(b) all surplus explosives are in a safe place; and

(c) if a number of persons are present, an agreed signal has been given by the person in charge.

23.4.2. Persons should be prevented from entering the firing area.

23.4.3. All shots should be fired as soon as practicable after loading.

23.4.4. All shots that could affect one another should be fired together, or in quick succession.

23.4.5. All shots in the same round should be fired together, provided that this does not preclude delayed blasting.

23.4.6. When necessary to prevent danger, the surface to be blasted should be covered by a blasting mat, mattress, sandbags, fascines or other suitable means to arrest flying fragments.

23.4.7. Only electric igniters or detonating fuse should be used in boreholes over 3 m (10 ft) deep.

23.5. Firing with fuse

23.5.1. The covering of fuse should be protected against damage.

23.5.2. In cold weather precautions should be taken to avoid cracking the waterproofing of the fuse.

23.5.3. Fuses should be long enough to enable the person firing to reach a place of safety.

23.5.4. To ensure that the end of the fuse is dry a short piece should be cut off.

23.5.5. Inserted fuses should not be twisted.

23.5.6. In underground operations fuses should only be lit with fuse-lighters specially designed for the purpose.

23.5.7. Explosives should not be held in the hand while fuses are being lit.

23.5.8. Fuses for firing powder explosives in boreholes should not glow or emit sparks.

23.5.9. The time element determined by the length of the fuse should not be interfered with in any manner, and the burning rate of fuses should not be artificially regulated.

23.5.10. At the time of ignition, the fuse should be free of loops.

23.6. Firing with electricity

23.6.1. Only electric firing devices approved by the competent authority should be used for electric blasting.

23.6.2. Blasting conductors should not be used for any other purpose.

23.6.3. Blasting conductors should not be laid in the same conduit as other conductors.

23.6.4. Detonator wires should not be uncoiled or shots fired by electric igniters:

(a) during dust storms, electric storms or in the presence of other sources of large charges of static electricity; or

(b) in the vicinity of radio-frequency transmitters.

23.6.5. Shots should not be fired with electricity when there is danger from stray currents.

23.6.6. The firing circuit should be kept completely insulated from the ground, other conductors, such as bare wires, rails and pipes, and paths of stray currents.

23.6.7. Before a round is fired the firing circuit should be tested for continuity and resistance.

23.6.8. If an electrical firing circuit has been found by test to be defective, it should not be used unless it can be safely and effectively repaired.

23.6.9. Attempts should not be made to fire rounds of shots with less current than that specified by the manufacturer.

23.6.10. All wire ends to be connected should be bright and clean.

23.6.11. Electric blasting cap wires should be kept disconnected from the power source and short-circuited until the moment of firing.

23.6.12. Only electric blasting caps of the same manufacture should be used in firing one circuit.

23.6.13. When firing by electricity from the mains:

(a) the voltage should not exceed 250 V;

(b) the firing circuit should not carry current for any other purpose;

(c) the firing circuit should not cross any other live conductors;

(d) the firing circuit should include an operating switch interlocked with a safety switch; and

(e) both switches should be of the locking, double pole, double throw type.

23.6.14. All radio transmissions should be restricted when firing with electricity.

23.7. After firing

23.7.1. No person should return to the firing area until the responsible person has given a clear signal indicating that:

(a) the smoke and fumes have dispersed; and

(b) the blaster has carefully examined it for misfires, partly exploded shots, unused explosives and other sources of danger, and found it safe.

23.7.2. After a blast has been fired by electricity from the mains:

(a) the operating switch and the safety switch should be locked;

(b) the blaster should keep the firing keys.

23.7.3.   After a blast has been fired, loose pieces of rock should be scaled from the firing area.

23.7.4.   When firing with fuses, if a misfire has occurred or is suspected no person should return to the firing area for at least one hour or such period of time left to the discretion of the responsible person authorised to give the signal to workers to return to the firing area.

23.7.5.   Where a misfire has occurred the unexploded charge should be blown out by a shot in a parallel borehole.

23.7.6.   If a misfire occurs in a well hole the blaster should decide whether to insert a fresh charge, remove the tamping, or take other measures to prevent danger.

23.8. Blasting with liquid oxygen

23.8.1.   Liquid oxygen should only be used as an explosive by workers who are familiar with its use and in accordance with instructions drawn up by the employer.

23.8.2.   Liquid-oxygen cartridges should be of a type approved by the competent authority.

23.8.3.   Instructions should be drawn up by the employer in order to:

(a) limit the total amount of liquid oxygen allowed on the work-site;

(b) specify the minimum distance that any liquid oxygen must be kept from the firing area;

(c) specify the maximum amount of liquid oxygen to be put into any one cartridge;

(d) specify the maximum weight of charged cartridges to be carried by any one worker; and

(e) specify the conditions in which containers of liquid oxygen may be emptied.

23.8.4.   Absorbent cartridges should not be made underground.

23.8.5. Gloves and clothes of workers handling cartridges should not be used for any other work, and should be kept free from grease.

23.8.6. Liquid oxygen should only be kept and transported in special containers not used for any other purpose.

23.8.7. Vehicles transporting liquid oxygen should not at the same time transport workers, explosives, detonators and other ignition devices, combustible materials, or greasy materials.

23.8.8. If a truck transporting liquid oxygen is attached to a train transporting workers it should not be placed next to a truck transporting workers.

23.8.9. Near containers of liquid oxygen no person should smoke, have an open light or handle calcium carbide or greasy objects or materials.

23.8.10. At the end of the work day containers of liquid oxygen should be emptied in accordance with instructions given by a competent person.

23.8.11. Made-up absorbent cartridges should only be impregnated in the firing area and when they are required for use.

23.8.12. Cartridges attached to fuses should not be impregnated.

23.8.13. For blasting with liquid oxygen, only special fuses should be used that do not burn faster in oxygen.

23.8.14. When blasting with fuse, the primed cartridge should be next to the tamping.

23.8.15. No blasting should be done with any cartridge whose useful life has finished.

23.8.16. No worker should return to the firing area for at least one hour after a blast:

(a) when relays have been used;
(b) when a large round has been fired; and
(c) when it has not been possible to hear all the shots distinctly.

23.9. Well-hole blasting

23.9.1.   Blasting of well-holes should be governed by instructions issued by the employer.

23.9.2.   If cartridge explosives are used:

(a) the hole should afford sufficient clearance to prevent cartridges from jamming;

(b) the cartridges should have a strong envelope;

(c) for explosives sensitives to damp, the envelope should be waterproof; and

(d) only cartridges of explosives authorised for the purpose by the competent authority should be dropped freely into well-holes.

23.9.3.   If liquid oxygen is used:

(a) the cartridges should be of a type to be approved by the competent authority;

(b) cartridges should be lowered carefully into the well-hole without being compressed;

(c) fittings on cartridges for attaching them to a cord, etc. should be of brass;

(d) the tamping should be of incombustible powdery material; and

(e) misfired shots should not be untamped.

23.9.4.   Shots should only be lighted by detonating fuse.

23.9.5.   Before they are loaded holes should be inspected for obstructions.

23.9.6.   Detonating fuses should, as far as practicable, be in one piece.

23.9.7.   If joins have to be made in detonating fuses they should not lie in the tamping.

23.10. Blasting in pockets or fissures

23.10.1.   When blasting in pockets or fissures:

(a) only electrical igniters should be used with black powder;

(b) electric blasting caps or detonating fuses should be used with other explosives and blasting agents;

(c) chlorate and perchlorate explosives should be used in cartridge form;

(d) where practicable, suitable funnels should be used for introducing loose explosive; and

(e) in fissured rock, cartridges should be introduced by means of non-sparking tubes or gutters.

23.10.2.   If necessary to prevent danger in neighbouring occupied places, blasting mats or other means should be used.

23.10.3.   Rounds should not be fired when blasting in pockets.

23.10.4.   Before every blast the pocket should be cleared of loose stones.

23.10.5.   If black powder or granulated ammonium nitrate is fired in a pocket it should be loaded with a copper or plastic funnel of adequate length.

23.10.6.   The hole leading to the pocket should slope steeply downward.

23.10.7.   No powder should be left adhering to the sides of the hole.

23.10.8.   After a blast:

(a) the pocket or fissure should not be explored or reloaded for at least one hour; and

(b) the pocket or fissure should be adequately cooled, if practicable by compressed air or by flooding with water.

24. Pile driving

24.1. General provisions

24.1.1. Pile drivers should be firmly supported on heavy timber sills, concrete beds or other secure foundation.

24.1.2. If necessary to prevent danger, pile drivers should be adequately guyed.

24.1.3. Precautions complying with the requirements of section 17.8 should be taken if it is necessary to erect pile drivers in dangerous proximity to electrical conductors.

24.1.4. If two pile drivers are erected at one place they should be separated by a distance at least equal to the longest leg.

24.1.5. Access to working platforms and the top pulley should be provided by ladders complying with the relevant requirements of Chapter 4.

24.1.6. Working platforms and operator's working areas should be adequately protected against the weather.

24.1.7. Winches used with pile drivers should comply with the requirements of section 5.11.

24.1.8. When leads have to be inclined:

(a) they should be adequately counterbalanced; and
(b) the tilting device should be secured against slipping.

24.1.9. Steam or air lines should consist of armoured hose or its equivalent.

24.1.10. Couplings of sections of hose should be additionally secured by ropes or chains.

24.1.11. The hose of steam and air hammers should be securely lashed to the hammer so as to prevent it from whipping if a connection breaks.

24.1.12. Steam and air lines should be controlled by easily accessible shut-off valves.

24.1.13.   Sheaves on pile drivers should be so guarded that workers cannot be drawn into them.

24.1.14.   Adequate precautions should be taken to prevent a pile driver from overturning.

24.1.15.   Adequate precautions should be taken by providing stirrups or by other effective means, to prevent the rope from coming out of the top pulley or wheel.

24.1.16.   Adequate precautions should be taken to prevent the hammer from missing the pile.

24.1.17.   If necessary to prevent danger, long piles and heavy sheet piling should be secured against falling.

24.2. Inspection and maintenance of pile-driving equipment

24.2.1.   No pile-driving equipment should be taken into use until it has been inspected and found to be safe.

24.2.2.   Pile-driving equipment in use should be inspected at suitable intervals.

24.2.3.   Pile lines and pulley blocks should be inspected before the beginning of each shift.

24.2.4.   Defective parts of pile drivers such as sheaves, mechanisms, slings and hose should only be repaired by competent persons.

24.2.5.   No steam or air equipment should be repaired while it is in operation or under pressure.

24.3. Operation of pile-driving equipment

24.3.1.   Only competent persons should operate pile drivers.

24.3.2.   Pile-driving operations should be governed by signals in conformity with the requirements of section 1.8.

24.3.3.   Workers employed in the vicinity of pile drivers should wear safety helmets or hard hats.

24.3.4. As far as practicable, piles should be prepared at a distance at least equal to twice the length of the longest pile from the pile driver.

24.3.5. Piles being hoisted in the rig should be so slung that they do not have to be swung round, and cannot inadvertently swing or whip round.

24.3.6. While a pile is being hoisted all workers not actually engaged in the operation should keep at a safe distance.

24.3.7. A hand rope should be fastened to a pile that is being hoisted, to control its movement.

24.3.8. Before a wood pile is hoisted into position it should be provided with an iron ring or cap over the driving end, to prevent brooming.

24.3.9. While a pile is being guided into position in the leads, workers should not put their hands or arms between the pile and the inside guide or on top of the pile, but should use a rope for guiding.

24.3.10. When treated wood piles are being driven, adequate precautions to prevent injury should be taken, such as the provision of personal protective equipment and barrier creams to prevent workers receiving eye or skin injuries from splashes of creosote.

24.3.11. When piles are driven at an inclination to the vertical, if necessary to prevent danger, they should rest in a guide.

24.3.12. No steam or air line should be blown down until all workers are at a safe distance.

24.3.13. Loose fuel containers such as barrels and drums should be kept in a safe place and adequately secured.

24.3.14. When a pile driver is not in use, the hammer should be blocked at the bottom of the leads.

24.4 Floating pile drivers

24.4.1. When pile drivers are working over water, precautions should be taken as required in paragraph 2.6.19 and in Chapter 19, and in particular a suitable boat should be kept readily available at all times.

24.4.2. All members of floating pile-driver crews should be trained to handle boats.

24.4.3. Floating pile drivers should be provided with a whistle, siren, horn or other effective signalling equipment.

24.4.4. Signalling operations should comply with the requirements of section 1.8.

24.4.5. If practicable, gangways complying with the requirements of section 3.3 should be provided for boarding floating pile drivers.

24.4.6. Floating pile drivers should be provided with adequate fire-fighting equipment.

24.4.7. The weight of machinery on a floating pile driver should be evenly distributed so that the deck of the installation is horizontal.

24.4.8. Steel pile-driver hulls should be divided into watertight compartments.

24.4.9. Watertight compartments should be provided with siphons for the removal of water seepage.

24.4.10. Deck hatches should have firmly fastened covers that fit flush with the deck.

24.4.11. Open hatches should be adequately fenced or guarded.

24.4.12. Fuel tanks below deck should be vented to the outside air.

24.4.13. Vents referred to in paragraph 24.4.12 should be provided with flame arrestors.

24.4.14. For each fuel tank below deck there should be an emergency shut-off valve on deck.

24.4.15. Adequate precautions should be taken with respect to any electrical equipment below deck, to avoid explosions.

24.4.16. Sufficient sheaves should be provided on deck to enable the pile driver to be safely manœuvred in any direction and safely secured in position.

24.4.17. The operator's cabin should afford an unrestricted view of the operations.

24.5. Sheet piling

24.5.1. If necessary to prevent danger from wind or other sources, a hand rope should be used to control the movement of steel sheet sections that are being transported.

24.5.2. Workers who have to sit on a steel sheet section to interlock sheets should be provided with stirrups or other devices to afford them a safe seat.

24.5.3. Workers handling sheets should wear gloves.

24.5.4. If necessary to prevent danger from displacement by the action of water, steel sheet sections should be braced until they are firmly in position.

24.5.5. If necessary to prevent danger from undercutting of the cofferdam by the action of water, a substantial berm should be installed upstream.

24.5.6. When workers are assembling sheet piling in water:

(a) precautions as required in paragraph 2.6.19 and in Chapter 19 should be taken;

(b) working platforms complying with the requirements of section 3.2 should be provided; and

(c) safety belts complying with the relevant requirements of Chapter 36 should be provided and worn.

24.5.7. Workers should not stand on sheet piling while it is being released from the slings, lowered or moved into position.

24.5.8. While it is being weighted with stones, etc., sheet piling should be securely moored.

24.5.9. Adequate pumping facilities should be available at cofferdams to keep them clear of water.

24.5.10. Adequate means of escape such as ladders and boats should be provided at cofferdams for the protection of workers in case of flooding.

24.5.11. Adequate supplies of life-saving equipment should be provided for workers employed on cofferdams, in particular as required in paragraph 2.6.19.

24.5.12. When sheet sections are being removed, their movements should be controlled by cables or other effective means.

24.5.13. When sheet piling is hauled over ice:

(a) the ice should have been tested and found strong enough to bear the maximum loads that it will have to carry;

(b) a safe walkway should be constructed along gaps made in the ice for lowering the piling;

(c) staging erected on the ice for handling materials should rest on grids;

(d) passageways for workers and vehicles should be suitably signposted;

(e) the consistency of the ice should be constantly checked;

(f) loads on the ice should be suitably reduced in the event of a thaw;

(g) above the piling at the workplace, a working platform complying with the requirements of section 3.2 should be installed; and

(h) rescue equipment such as gaffs and ring buoys should be kept available.

25. Concrete work

25.1. General provisions

25.1.1. The construction of heavy reinforced concrete roofs and other heavy overhead structures should be based on plans that:

(a) include specifications of the steel and concrete and other material to be used, including technical methods for safe placing and handling;

(b) indicate the type, strength and arrangement of load-bearing parts; and

(c) provide, if appropriate, calculations of the strength of roofs or other heavy overhead structures made with prefabricated parts.

25.1.2. During the construction of heavy reinforced concrete roofs and other heavy overhead structures, a daily record should be kept of the progress of the work, including indications of all data which could affect the curing of the concrete.

25.2. Preparation, pouring and erection of concrete

25.2.1. Workers handling cement and concrete should:

(a) wear close-fitting clothing, gloves, helmets or hard hats, safety goggles and suitable footwear and, if necessary to prevent danger, respirators or masks;

(b) keep as much of the body covered as is practicable;

(c) take all practicable precautions to keep cement and concrete away from the skin; and

(d) wash frequently and, if necessary, use appropriate cream on exposed skin.

25.2.2. When work is done with cement, lime and other dusty materials, or machines for crushing, grinding or sifting building materials are operated in enclosed premises:

(a) the premises should be provided with adequate general and exhaust ventilation; or

(b) other precautions should be taken to prevent the dispersal of dust.

25.2.3. Wherever work is done with unslaked lime, precautions should be taken to prevent the dispersal of dust.

25.2.4. Where workers are handling unslaked lime, precautions should be taken to prevent risks of burns.

25.2.5. The controls of machines processing cement, lime and other dusty materials should be in a dust-free place.

25.2.6. Lime pits should be fenced or enclosed.

25.2.7. Lime pits should be filled and emptied by devices that do not require workers to go into the pit.

25.2.8. Elevators, hoists, screens, bunkers, chutes and other equipment for storing, transporting and working ingredients of concrete, grouting, etc. should be securely fenced to avoid contact with moving parts that are not safe by position.

25.2.9. Screw conveyors for cement, lime and other dusty materials should be completely enclosed.

25.2.10. Blocked conveyors should be stopped before attempting to clear the blockage.

Buckets

25.2.11. Concrete buckets for use with cranes and aerial cableways should be free from projections from which accumulations of concrete could fall.

25.2.12. Movements of concrete buckets should be governed by signals in conformity with the requirements of section 1.8.

25.2.13. Loaded concrete buckets should be guided into position by appropriate means.

25.2.14. Concrete buckets positioned by crane or aerial cableways should be suspended by safety hooks.

Pipes and pumps

25.2.15. Scaffolding carrying a pipe for pumped concrete should be strong enough to support the pipe when filled and all the workers who may be on the scaffold at the same time, with a safety factor of at least 4.

25.2.16. Pipes for carrying pumped concrete should:

(a) be securely anchored at the ends and at curves;

(b) be provided near the top with air release valves; and

(c) be securely attached to the pump nozzle by a bolted collar or equivalent means.

25.2.17. While concrete pump pipes are being cleaned with water or compressed air, they should not be coupled or uncoupled.

25.2.18. While pipes are being blown out with compressed air, all workers not taking part in the operation should be kept at a safe distance.

25.2.19. The operation of concrete pumps should be governed by signals complying with the requirements of section 1.8.

25.2.20. Pressure gauges on pumps should be checked at the beginning of every shift.

25.2.21. Workers employed around a concrete pump should wear safety goggles.

Mixing and pouring concrete

25.2.22. Concrete should not contain any material that would affect its setting, weaken it or attack steel.

25.2.23. When dry ingredients of concrete are being mixed in confined spaces such as silos:

(a) the dust should be exhausted; or

(b) if the dust cannot be exhausted, the workers should wear respirators.

25.2.24. Frozen building materials should not be used.

25.2.25. Concrete should be used immediately after continuous mixing.

25.2.26. During pouring, shuttering and its supports should be continuously watched for defects.

25.2.27. When concrete is being tipped from buckets, workers should keep out of range of any kickback due to concrete sticking to the bucket.

25.2.28. While concrete is beginning to set, it should be protected against extremes of temperature, running water, chemical agents and jolting.

25.2.29. Loads should not be dumped or placed on setting concrete.

Concrete panels and slabs

25.2.30. All parts of a panel or slab should be hoisted uniformly.

25.2.31. Panels should be adequately braced in their final position and the bracing should remain in place until the panel is adequately supported by other parts of the construction.

25.2.32. When concrete slabs are lifted on to columns by jacks:

(a) a metal scaffold tower equipped with a lifting appliance should be provided;

(b) the working platform should be about 1 m (3 ft 3 in) below the top of the column;

(c) if a hoist is mounted on the scaffold, it should allow adequate clearance to manœuvre the collars over the columns; and

(d) access to the working platform should be by means of ladders built in or attached to the scaffold.

25.2.33. During lifting operations with jacks:

(a) collar clearances on columns should be kept free from obstructions;

(b) openings in slabs should be suitably covered or barricaded;

(c) slabs should be kept horizontal;

(d) if slabs have to be placed temporarily on the columns, as in tall structures, adequate shoring or other supports should be used;

(e) a jack should not be released at a column before the temporary support is complete;

(f) adequate precautions should be taken to keep the columns vertical; and

(g) all column connections should be made and found to be safe by inspection before a panel or slab raised by a hoisting appliance is released from the appliance.

25.2.34. Temporary bracing should be securely fastened to prevent parts from falling when panels are being moved.

25.2.35. When concrete slabs are tilted into position:

(a) the slabs should be strong enough to be self-supporting; or

(b) the slabs should be adequately braced while being lifted.

Stressed and tensioned elements

25.2.36. Workers should not stand directly over jacking equipment while stressing is being done.

25.2.37. Pre-stressed concrete units should only be handled at points on the unit and by devices specified by the manufacturer.

25.2.38. During transport, pre-stressed girders and beams should be kept upright by bracing or other effective means.

25.2.39. Anchor fittings for pre-tensioned strands should be kept in a safe condition in accordance with the manufacturer's instructions.

25.2.40. When any tensioning element is being jacked, the anchor should be kept turned up close to the anchor plate if practicable, so as to reduce impact and damage if a hydraulic line fails.

25.2.41. Workers should not stand behind jacks or in line with tensioning elements and jacking equipment during tensioning operations.

25.2.42. Workers should not cut wires under tension before the concrete is adequately hardened.

Vibrators

25.2.43. Only workers in good physical condition should operate vibrators.

25.2.44. All practical measures should be taken to reduce the amount of vibration transmitted to vibrator operators.

25.2.45. Pipes delivering concrete to vibrators should:

(a) have the couplings between sections secured with safety chains or other effective means; and

(b) have the delivery end secured against displacement.

25.2.46. When electric vibrators are used:

(a) they should be earthed;

(b) the leads should be heavily insulated; and

(c) the current should be switched off when they are not in use.

25.3. Steel reinforcement

25.3.1. Where necessary to prevent danger, protection should be provided for reinforcing rods projecting from flooring or from walls.

25.3.2. When lashing reinforcing rods in structural parts such as walls and pillars, projecting ends should not be left so as to cause danger.

25.3.3. Reinforcing rods should not be so stored on scaffolding or shuttering as to threaten its stability.

25.4. Concrete-bucket towers

25.4.1. Towers and masts with pouring gutters or conveyor belts should:

(a) be erected by competent persons; and

(b) be inspected daily.

25.4.2. Concrete-bucket towers should be adequately guyed.

Enclosure

25.4.3.   A concrete-bucket tower inside a structure, near a scaffold or edge of a shaftway or floor opening, should be enclosed on all sides with heavy wire netting, wood slats or equivalent material.

25.4.4.   The enclosure should extend at least 2.40 m (8 ft) above the adjacent scaffold or floor.

25.4.5.   A concrete-bucket tower outside a structure should be similarly enclosed to a height of at least 2.40 m (8 ft) from the bottom.

Gangways

25.4.6.   At each floor level there should be a gangway leading to the tower protected by guard-rails and toe-boards complying with the requirements of paragraphs 2.6.1 to 2.6.5.

Working platforms

25.4.7.   A substantial working platform should be constructed at the point where the concrete is dumped into the chute.

25.4.8.   The platform should be provided with:

(a) an access ladder complying with the relevant requirements of Chapter 4; and

(b) guard-rails and toe-boards complying with the requirements of paragraphs 2.6.1 to 2.6.5.

Chutes

25.4.9.   If a bucket discharges into a chute, the chute should:

(a) be substantially constructed of metal or wood or equivalent material;

(b) be substantially supported; and

(c) extend from the tower to the point where the concrete is to be poured or transferred to vehicles or hoppers.

Spouts (gutters)

25.4.10.   Spouts (gutters) should be adequately guyed to prevent displacement by wind or other causes.

25.4.11.   Spouts (gutters) should be flushed out at the end of each run of concrete.

25.4.12.   Areas below spouts (gutters) should be fenced off as far as practicable to prevent danger to workers from falling concrete.

25.4.13.   Spouts (gutters) should be so made that as far as practicable:

(a) they cannot be stopped up; and
(b) stones are not flung out.

25.4.14.   Outrigger and flying spouts (gutters) should be equipped with a handrope or other secure hand-hold for workers attaching the supporting lines.

25.4.15.   Workers who have to go on spouts (gutters) should wear a safety belt complying with the relevant requirements of Chapter 36.

Hoisting equipment

25.4.16.   The winch should be so placed that the operator can see the filling, hoisting, emptying and lowering of the bucket.

25.4.17.   If the winch operator cannot watch the bucket, he should, where practicable, be provided with an adequate means indicating its position.

25.4.18.   Guides for the bucket should be correctly aligned and so maintained as to prevent the bucket from jamming in the tower.

25.4.19.   The overhead sheaves should be firmly supported by beams mounted on a framework or by other effective means.

25.4.20.   Workers should not ride in the bucket except for necessary inspection and maintenance purposes, and then adequate measures should be taken for their safety.

Silos

25.4.21.   No worker should enter the silo or the storage bin on the tower to make adjustments or for other purposes unless the power has been cut off and effective steps have been taken to prevent it from being switched on again.

Sumps

25.4.22.   The tower should have a pit or sump deep enough to hold any spillage from the bucket.

25.4.23.   The pit or sump referred to in paragraph 25.4.22 should be drained.

25.4.24.   No worker should be employed in the pit unless the bucket has been securely blocked on supports at a safe height.

Track-mounted towers

25.4.25.   If towers or masts are mounted on wheels, the load should be taken off the wheels and axles while they are in operation.

25.4.26.   Guard-rails and toe-boards for towers or masts should comply with the requirements of paragraphs 2.6.1 to 2.6.5.

25.5. Form work

Placing of shuttering (forms)

25.5.1.   Shuttering should be examined, erected and dismantled under the supervision of qualified and experienced persons and, as far as practicable, by workers familiar with the work.

25.5.2.   The necessary information for the erection of shuttering, including particulars of the spacing of stringers and props and the fastening of props to stringers, should be provided for the workers in the form of sketches or scale drawings.

25.5.3.   Lumber and supports for shuttering (forms) should be adequate, having regard to the loads to be borne, spans, setting temperature and rate of pour.

25.5.4.   Material for shuttering should be carefully examined before being used.

25.5.5.   Forms and form panels should be provided with adequate U-bolts or other lifting attachments.

25.5.6. Heavy building materials should not be dumped or stored on shuttering.

25.5.7. Where necessary to prevent danger, scaffolding should be used:

(a) for positioning and fastening panels or slabs; and

(b) when high ceilings and arches, or heavy structural parts, are being constructed.

25.5.8. Shuttering should be adequately braced horizontally and diagonally in both longitudinal and transverse directions.

Shoring (supports)

25.5.9. Shoring (supports) should be strong enough to withstand safely:

(a) vertical loads imposed by the shuttering, concrete, appliances, impacts, vibrations, etc.; and

(b) lateral thrusts from supports or neighbouring operations.

25.5.10. Shores and props should:

(a) be suitably spaced;

(b) provide adequate footing;

(c) be adequately secured in position; and

(d) be adequately braced horizontally and diagonally in both directions.

25.5.11. Props should be of steel or sound straight and straight-grained timber.

25.5.12. If wooden props with joins in them are used:

(a) the ratio of props with joins to props without joins should not be greater than 1 to 2;

(b) the props with joins should be distributed equally over the shuttering;

(c) no prop should have more than one join;

(d) the joins should be secured by straps in a manner adequate to prevent buckling of the props; and

(e) diagonal bracing should be provided at the level of each join.

25.5.13.   Shores and props should rest on sills of adequate dimensions or other suitable firm bases, taking into account the axial loads to be applied.

25.5.14.   Sills should rest on a firm foundation, and never on frozen ground or loose earth.

25.5.15.   Extensible props should be fitted with a limiting device to prevent them from being pulled or screwed to an excessive length.

25.5.16.   If possible, the maximum permissible length should be clearly marked on extensible props.

25.5.17.   All adjustable shoring should be locked in position when adjusted.

25.5.18.   Shoring should be so arranged that when it is being removed sufficient props can be left in place to afford support necessary to prevent danger.

25.5.19.   Shoring should be adequately protected from damage from moving vehicles, swinging loads, etc.

25.5.20.   Protruding nails, wires and other accessories should be removed from shoring or otherwise made safe.

25.5.21.   Shoring should be left in place until the concrete has acquired sufficient strength to support safely not only its own weight but also any imposed loads.

25.5.22.   Shoring should be adequately braced or tied together to prevent deformation or displacement.

Removal of shuttering

25.5.23.   No shuttering should be removed until the authorisation has been given by a competent person who has made sure that the concrete has sufficiently set to support itself and any superimposed loads.

25.5.24.   To prevent danger from falling parts when shuttering is being taken down, the shuttering should as far as practicable be taken down whole, or else remaining parts should be supported.

25.5.25. When shuttering is being removed, the props and panels should be lowered steadily and not be knocked off.

25.5.26. Frozen concrete should not be subjected to any load.

25.5.27. If necessary to prevent danger, workers stripping forms should wear safety belts complying with the relevant requirements of Chapter 36.

25.5.28. Shuttering material that is taken down should be so stored that it does not interfere with workplaces, passageways or transport ways.

Lifting appliances

25.5.29. Mechanical, hydraulic or pneumatic lifting appliances for handling forms should be provided with automatic holding devices to prevent danger if the lifting mechanism power fails.

25.5.30. Lifting appliances should be uniformly spaced and securely anchored.

25.5.31. Vacuum lifting appliances should only be applied to smooth clean surfaces.

25.5.32. Vacuum lifting devices should be provided with an automatic cut-off to prevent loss of suction in the event of a power or equipment failure.

Supporting of slabs and beams

25.5.33. Where necessary to prevent danger, adequate shoring should be provided to support slabs and beams as a protection against superimposed loads.

26. Roof work

26.1. General provisions

26.1.1.   Where necessary to prevent danger, precautions should be taken to prevent:

(a) workers employed on roofs from falling off them; and

(b) tools or other objects and materials from falling from roofs.

26.1.2.   Work on roofs should not be carried on in high winds, violent storms or heavy snowfalls or when the roof is covered with ice.

26.1.3.   Crawling boards should be at least 25 cm (10 in) wide.

26.1.4.   Crawling boards should be securely fastened to a firm structure.

26.1.5.   Roofing brackets should fit the slope of the roof.

26.1.6.   Roofing brackets should be securely supported.

26.2. Steep roofs

26.2.1.   Work on steeply sloping roofs should only be done by workers who are physically and psychologically fit for such work.

26.2.2.   When work is done on steeply sloping roofs, sufficient and suitable crawling boards or ladders should be provided, and firmly secured in position as soon as practicable.

26.2.3.   When work is done on steeply sloping roofs, if practicable, a working platform complying with the requirements of section 3.2 should be provided and firmly secured in position.

26.2.4.   If necessary to prevent danger when work is done on steeply sloping roofs, a parapet wall or adequate guard-rail and toe-board should be provided at the edge of the roof or a catch platform or adequate safety nets should be provided or other adequate precautions should be taken to prevent workers falling over the edge.

26.2.5. On all roofs with a slope of more than 20° the workers should be provided with a safety belt complying with the relevant requirements of Chapter 36 if they could fall more than 1.5 m (5 ft).

26.3. Fragile roofs

26.3.1. Where workers are employed on roofs of fragile material through which they could fall more than 1.5 m (5 ft), they should be provided with sufficient suitable ladders, duck ladders or crawling boards or other safe workplaces and walkways.

26.3.2. Ladders, duck ladders, crawling boards and walkways should be firmly secured in position.

26.3.3. Workers should not leave the workplaces, ladders, walkways, etc. provided to go on the roof.

26.3.4. When necessary to prevent danger, suitable material such as steel wire mesh should be placed in position before any roof sheeting of asbestos cement or other fragile material is placed.

26.3.5. Corrugated asbestos cement sheeting for roofs should be mature and free from dangerous cracks and other dangerous defects.

26.3.6. Purlins or other intermediate supports for corrugated asbestos cement sheeting should be sufficiently close together to prevent danger.

26.3.7. Parts of roofs that cannot be walked on safely should be provided with notices warning persons to keep off them.

27. Painting

27.1. General provisions

27.1.1. Paints, pigments, solvents, thinners, fillers, lacquers and similar materials containing substances that are explosive, flammable, toxic or otherwise dangerous should comply with the relevant provisions of Chapter 21 and this chapter.

27.1.2. As far as practicable, use should not be made of paints:

(a) containing lead, chrome, arsenic or other dangerous pigments or dyestuffs; or

(b) containing dangerous solvents such as benzene, dichlorethane, methanol (methylated spirit), carbon tetrachloride, and trichlorethylene.

27.1.3. No solvent, preservative, metal coating or other chemical material should be used unless any fire, toxic or other dangers that may occur in its handling or use are known to all concerned in the operations.

27.1.4. When not in use, containers or packages containing paints, varnishes, lacquers and other volatile painting materials should:

(a) be kept tightly closed; and

(b) kept away from sparks, flames, sources of heat and the sun's rays.

27.1.5. Painting materials should not be heated except in a water jacket at a moderate temperature or in a special device approved by the competent authority.

27.1.6. No open flame or electrical appliances with open heating elements should be used to dry paint or similar coating material containing a highly flammable or explosive solvent.

27.1.7. Dirty wiping rags, paint scrapings, paint-saturated debris, etc. should not be allowed to accumulate but should:

(a) be cleaned up at frequent intervals; or

(b) be kept in closed metal containers.

27.1.8.   No smoking, open flames or sources of ignition should be allowed in places where paint and other flammable substances are stored.

27.1.9.   Suitable fire extinguishers should be kept available at places where flammable paints are stored, mixed or used.

27.1.10.   Workers using paint other than water-mixed paint in a confined space such as a closed room, or using indoors paint that liberates harmful fumes, should be provided with:

(a) adequate natural or artificial ventilation; or
(b) suitable respirators.

27.1.11.   When necessary to protect the skin, painters should wear gloves and use appropriate barrier creams.

27.1.12.   When paint is removed by solvents containing benzene, carbon tetrachloride or other harmful substances, the workers should wear suitable respirators.

27.1.13.   When such work is done in a confined space, adequate exhaust ventilation should be provided and the workers should wear suitable respirators.

27.1.14.   Skin should be cleaned by suitable materials.

27.1.15.   If solvents are used to clean skin, they should be non-harmful.

27.1.16.   When paint is being sanded by machine:

(a) the sandpaper should be wet or the machine should be equipped with an adequate dust collector; and
(b) the worker should be provided with suitable protective clothing and a suitable respirator.

27.1.17.   Where work cannot be carried out from a scaffold, the painting of the outside of window frames, etc. should be done in compliance with the relevant requirements of section 29.11.

27.1.18.   Particular care should be exercised when painting near electrical installations where spark hazards, static or otherwise, exist.

27.1.19. No person should work or walk on, or move or manipulate, iron work or steel work covered with wet paint or cement wash except in so far as is necessary for painting or cement-washing.

27.2. Lead paint

27.2.1. Women and young persons should not be employed on work with lead paint.

27.2.2. Lead paint should not be used for painting buildings except in the form of paste or of paint ready for use.

27.2.3. Red lead in the raw or dry state should only be used for preparing stopping or filling material.

27.2.4. White lead or sulphate of lead should not be used or manipulated in the preparation of painters' stopping material except under an effective exhaust draught that removes the dust produced at or near its source.

27.2.5. Lead paint should only be stored in receptacles plainly marked to show the contents.

27.2.6. Lead paint should not be applied in the form of spray inside buildings.

27.2.7. No lead-painted surface other than that of iron or steel work should be rubbed down or scraped by a dry process.

27.2.8. No lead-painted surface of iron or steel work should be rubbed down or scraped by a dry sandpapering process.

27.2.9. All debris produced by rubbing down or scraping any lead-painted surface should be removed before it becomes dry.

27.2.10. Adequate washing facilities complying with the requirements of section 37.5 should be provided for persons employed in the painting of buildings and liable to come into contact with lead paint.

27.2.11. Personal clothing of workers should be suitably protected from being soiled by lead paint.

27.2.12. Persons regularly employed on the painting of buildings with lead paint should undergo pre-employment examination and periodical medical examinations.

27.2.13. Persons whose health has been or appears likely to be injuriously affected by painting with lead paint should not be further employed on such work.

27.2.14. All persons employed on painting with lead paint should be given adequate instructions in the precautions to be taken.

27.2.15. Persons employed in painting and liable to come into contact with lead paint should:

(a) wear head covering and overalls or other working clothes that completely cover them while they are working;
(b) wash their hands before eating, and before leaving work; and
(c) prevent paint from entering their mouths in any manner whatsoever.

27.2.16. Working clothes should not be worn at meal times.

27.2.17. Working clothes should be washed at least once a week.

27.3. Unsaturated polyesters

27.3.1. If unsaturated polyesters and organic peroxides are used they should be treated as highly flammable and explosive substances and the relevant precautions required by section 21.2 should be taken.

27.3.2. Containers of over 25 litres (5½ gallons) of organic peroxides should be stored in a special building or room.

27.3.3. Quantities up to 25 litres (5½ gallons) of organic peroxides should be kept in a cupboard ventilated by outside air.

27.3.4. A building, room or cupboard containing organic peroxides should:

(a) be of fire-resisting construction;
(b) have the roof designed as an explosion vent;

(c) be kept cool; and
(d) not be used for any other purpose.

27.3.5. Before beginning work, workers using polyesters should rub barrier creams on the hands.

27.3.6. Workers using synthetic resins should protect their hands as far as practicable, for instance by wearing suitable gloves.

27.3.7. Resins and hardeners (setters) should be prevented from getting into gloves.

27.3.8. Polyester spilled on the skin should be immediately wiped off with tissue paper, but not rags.

27.3.9. Only competent persons should handle organic peroxides.

27.3.10. In handling organic peroxides, persons should wear goggles or face screens.

27.3.11. Organic peroxides should not be mixed directly with accelerators or added to hot bases; the accelerators should first be mixed with the resin.

27.3.12. Precautions should be taken to prevent peroxides and polyesters from being spilled on floors and worktables.

27.3.13. Spilled liquid peroxide and polyesters should be cleaned up immediately with suitable absorbent material, not with a cloth.

27.3.14. Ignited solid peroxides should be extinguished with water and liquid peroxides with sand or other powder extinguishers.

27.3.15. Waste peroxide should be destroyed by a method that prevents explosions, for instance by dissolving it in sodium hydroxide.

27.3.16. Polyester resins should be mixed with peroxides, fillers, pigments, etc. in a place specially set aside for the purpose.

27.3.17. If harmful pigments and fillers, such as quartz, asbestos or other harmful powders are used, local exhaust ventilation should be provided.

27.3.18. Escaping styrene vapour should be exhausted as it arises.

27.3.19. Places, appliances and tools used for work with polyesters should be cleaned daily.

27.3.20. No flammable solvents should be used for cleaning operations specified in paragraph 27.3.19.

27.3.21. Containers of organic peroxides should be:

(a) kept closed;
(b) kept clean and protected against dirt; and
(c) distinctly identifiable.

27.3.22. When unsaturated polyesters and lacquers based on them are sprayed:

(a) the work should only be done in special booths in which no other material is sprayed;
(b) only centrifugal fans should be used to ventilate the booths; and
(c) residues from the booth should be burned in the open air, or kept in closed metal containers.

27.4. Spray painting

27.4.1. Spray painting should not be done with any material containing carbon bisulphide, carbon tetrachloride, arsenic, arsenic compounds or more than 1 per cent of benzene or methyl alcohol.

27.4.2. A sufficient number of fire extinguishers of the foam or other suitable type should be maintained at the place where any material having a nitro-cellulose or other flammable content is being sprayed.

27.4.3. After use, cotton waste, cleaning rags and the like should be kept in metal containers with close-fitting lids.

27.4.4. The containers referred to in paragraph 27.4.3 should be removed from the building or structure at the close of work each day.

27.4.5. No person should smoke, or have any fire, open flame, or other source of ignition in any place in a building or structure in which spray painting is being done with flammable solvent.

27.4.6. The air at any place in which spray painting is being done should be changed by natural ventilation, exhaust fan or other effective means at least 20 times in each hour.

27.4.7. If spray painting is done in a confined space, the workers employed in this space should be provided with suitable respirators unless the ventilation provided in the space is adequate to prevent danger.

27.4.8. Except when being mixed or sprayed, silica paint should be kept in receptacles clearly marked to show their contents.

27.4.9. Workers employed in spray painting should be provided with:

(a) overalls, head covering and gloves; and
(b) a sufficient quantity of material capable of removing paint or spraying mixture from the hands and face.

27.4.10. Workers employed in spraying silica paint should be provided with an adequate supply of hot water for washing, towels, nail brushes and soap.

27.4.11. Workers employed in outside spray painting with lead or silica paint or other harmful substance, or rubbing down or scraping a surface painted with such a substance, should be provided with suitable respirators.

27.4.12. Spray-gun operators should:

(a) adjust the atomisation pressure of the gun so as not to create excessive mist;
(b) use the gun so that neither they themselves nor any other workers remain between the gun and any ventilation fan; and
(c) not test the gun by indiscriminate or careless spraying.

27.5. Airless spray painting

27.5.1. In addition to complying with the requirements of section 4 of this chapter, airless spray painting should comply with the requirements of this section.

27.5.2. Hoses, guns and pressure vessels used for airless spray painting should have special high-pressure fittings not interchangeable with low-pressure fittings.

27.5.3. Guns should be equipped with:

(a) a guard over the trigger that prevents the gun from spraying if it is dropped or struck; and

(b) a safety latch that has to be unlocked before the gun can spray.

27.5.4. The safety latch should be kept in the non-operating position when the gun is not spraying.

27.5.5. When the gun is spraying with flammable material, the gun itself and the object being sprayed should be effectively earthed to prevent static sparking.

27.5.6. The gun should not be disconnected from the fluid hose or the hose from the pump until the pressure has been released.

27.5.7. The operator should:

(a) not touch the gun trigger while the gun is not spraying; and

(b) take any other precautions required to prevent the gun from discharging accidentally.

27.5.8. The gun should never be pointed at any person.

27.5.9. Guns should only be cleaned in accordance with the manufacturer's instructions.

28. Welding and flame cutting

28.1. General provisions

28.1.1. Welders should wear fire-resistant protective clothing and equipment such as fire-resistant gauntlets and aprons, helmets and goggles with suitable filter lenses.

28.1.2. Welders should wear clothing that is free from grease, oil and other flammable material.

28.1.3. Workers removing excess metal, slag, etc. should:

(a) wear gloves and goggles or a face screen;

(b) chip away from the body; and

(c) ensure that other persons are not struck by chips.

28.1.4. Adequate precautions should be taken to protect persons working or passing near welding operations from dangerous sparks and radiation.

28.1.5. When welding or cutting is being done on materials containing toxic or harmful substances or liable to produce toxic or harmful fumes, adequate precautions should be taken to protect workers from the fumes either:

(a) by exhaust ventilation; or

(b) by respiratory protective equipment.

28.1.6. At places where welding machines are operated by internal combustion engines, adequate ventilation should be provided.

28.1.7. A suitable fire extinguisher should be kept ready for immediate use at places where welding is done.

28.1.8. The oxygen pressure for welding should always be high enough to prevent acetylene flowing back into the oxygen cylinder.

28.1.9. Acetylene should not be used for welding at a pressure exceeding 1 atm. gauge.

28.1.10. Adequate precautions should be taken to prevent:

(a) fires being started by sparks, slag or hot metal; and

(b) damage to fibre ropes from heat, sparks, slag or hot metal.

28.1.11. Precautions should be taken to prevent flammable vapours and substances from entering the working area.

28.1.12. Underwater welding and flame cutting operations should comply with the requirements of paragraphs 34.4.31 to 34.4.40.

28.1.13. For cylinders containing gas, the requirements of section 18.4 should be observed.

Welding at places with fire risks

28.1.14. Unless adequate precautions are taken, no welding or cutting operations should be allowed near places where combustible materials are stored, or near materials or plant where explosive or flammable dusts, gases or vapours are likely to be present or given off.

28.1.15. Combustible materials and structures that cannot be removed from the vicinity of welding operations should be shielded by asbestos or protected by other suitable means.

Welding in confined spaces

28.1.16. When welding and cutting operations are being carried out in a confined space:

(a) adequate ventilation, by means of exhaust fans or forced draught as the conditions require, should be constantly provided; oxygen should not be used for this purpose;

(b) no blowpipe should be left unattended inside a tank or vessel or other confined space during meals or other interruptions of the work;

(c) the workers should take all necessary precautions to prevent unburned combustible gas or oxygen from escaping inside a tank or vessel or other confined space; and

(d) when necessary to prevent danger, an attendant should watch the welder or welders from outside.

Welding on containers for explosive or flammable substances

28.1.17.   Welding or cutting operations on containers in which there are explosive or flammable substances should not be allowed.

28.1.18.   Welding or cutting operations on any container that has held explosive or flammable substances, or in which flammable gases may have been generated, should only be undertaken:

(a) after the container has been:

  (i) thoroughly cleansed by steam or other effective means; and
  (ii) found by air tests to be completely free from combustible gases and vapours; or

(b) after the air in the container has been completely replaced by an inert gas or by water.

28.1.19.   If an inert gas is used as laid down in paragraph 28.1.18, after the vessel has been filled with gas, the gas should continue to flow slowly into it throughout the welding or cutting operation.

28.1.20.   Before starting any welding operations on, or otherwise applying heat to, closed or jacketed containers or other hollow parts, such containers or parts should be adequately vented in a suitable manner.

Gas cylinders

28.1.21.   Gas cylinders should be inspected, stored, handled and transported in conformity with the requirements of section 18.4.

28.1.22.   When in use, cylinders should be held in an upright position by straps, collars or chains.

28.1.23.   Devices referred to in paragraph 28.1.22 should be such that the cylinders can be rapidly removed in an emergency.

28.1.24.   Welders should not tamper with or attempt to repair safety devices and valves on gas cylinders.

28.1.25. When acetylene cylinders are coupled, flash arrestors should be inserted between the cylinder and the coupler block, or between the coupler block and the regulator.

28.1.26. Only acetylene cylinders of approximately equal pressure should be coupled.

28.1.27. Gas should not be taken from a cylinder unless a pressure-reducing regulator has been attached to the valve.

28.1.28. Only the right pressure-reducing regulator should be used for the gas in the cylinder.

28.1.29. Cylinder valves should be kept free from grease, oil, dust and dirt.

28.1.30. The emptying of leaky cylinders, referred to in paragraph 18.4.14, should be carried out slowly.

Hose

28.1.31. Only hose specially designed for welding and cutting operations should be used to connect an oxy-acetylene torch to gas outlets.

28.1.32. Hose lines for oxygen and for acetylene should be of different colours and preferably of different sizes.

28.1.33. Hose connections should be sufficiently tight to withstand without leakage a pressure twice the maximum delivery pressure of the pressure regulators in the system.

28.1.34. Care should be taken that hose does not become kinked or tangled, stepped on, run over or otherwise damaged.

28.1.35. Any length of hose in which a flashback has burned should be discarded.

28.1.36. No hose with more than one gas passage should be used.

28.1.37. Only soapy water should be used for testing hose for leaks.

Torches

28.1.38. When torches are being changed the gases should be shut off at the pressure-reducing regulators and not by crimping hose.

28.1.39. Torches should be lit with friction lighters, stationary pilot flames or other safe source but not with matches.

28.2. Electric welding

Equipment

28.2.1. Welding machines should be controlled by a switch mounted on or near the machine framework that, when opened, immediately cuts off the power from all conductors supplying the machine.

28.2.2. Welding circuits should be so designed as to prevent the transmission of high potential from the source of supply to the welding electrodes.

28.2.3. The maximum open circuit voltage should be in accordance with national or international codes or standards.

28.2.4. Electrode conductors or cables should not be excessive in length and should not be longer than necessary to perform the work.

28.2.5. Return conductors should be taken directly to the work and securely connected mechanically and electrically to it or to the work bench, floor, etc. and to an adjacent metallic object.

28.2.6. Cables should be supported so as not to create dangerous obstructions.

28.2.7. Motor generators, rectifiers and transformers in arc-welding or cutting machines, and all current-carrying parts, should be protected against accidental contact with uninsulated live parts.

28.2.8. Ventilating slots in transformer enclosures should be so designed that no live part is accessible through any slot.

28.2.9. Frames of arc-welding machines should be effectively earthed.

28.2.10. In hand-operated arc-welding machines, cables and cable connectors used in arc-welding circuits should be effectively insulated on the supply side.

28.2.11. The outer surface of electrode holders of hand-operated arc-welding machines, including the jaw so far as practicable, should be effectively insulated.

28.2.12. Electrode holders of hand-operated arc-welding machines should, if practicable, be provided with discs or shields to protect the operators' hands from the heat of the arcs.

28.2.13. Only heavy-duty cable with unbroken insulation should be used.

28.2.14. Circuit connections should be waterproof.

28.2.15. When lengths of cable have to be joined, only insulated connectors should be used, on both the earth line and the electrode holder line.

28.2.16. Connections to welding terminals should be made at distribution boxes, socket outlets, etc. by bolted joints.

28.2.17. Welding terminals should be adequately protected against accidental contact by enclosures, covers or other effective means.

28.2.18. Electrode holders should:

(a) have adequate current capacity; and

(b) be adequately insulated to prevent shock, short circuiting or flashovers.

Operations

28.2.19. Arc-welding and cutting operations that are carried on at places where persons other than the welders are working or passing should be enclosed by means of suitable stationary or mobile screens.

28.2.20. Walls and screens of both permanent and temporary protective enclosures should absorb harmful rays from the welding equipment and prevent reflection, and if necessary be painted or otherwise treated for the purpose.

28.2.21. When arc-welding is done in damp confined spaces:

(a) the electrode holders should be completely insulated; and

(b) the welding machine should be outside the confined space.

28.2.22. Welders should take adequate precautions:

(a) to prevent any part of their body from completing an electric circuit;

(b) to prevent contact between any part of the body and the exposed part of the electrode or electrode holder when they are in contact with metal; and

(c) to prevent wet or damaged clothing, gloves and boots from touching any live part.

28.2.23. Welding circuits should be switched off when not in use.

28.2.24. Electrodes should only be inserted in the holder with insulating means such as insulating gloves.

28.2.25. Electrode and return leads should be adequately protected against damage.

28.2.26. Live parts of electrode holders should be inaccessible when they are not in use.

28.2.27. Live parts of electrode holders should not be allowed to come into contact with metallic objects when not in use.

28.2.28. Electric arc-welding equipment should not be left unattended with the current switched on.

29. Other building operations

29.1. Erection of prefabricated parts

General provisions

29.1.1. As far as practicable the safety of prefabricated parts should be ensured by appropriate means, such as provision and use of:

(a) ladders;
(b) gangways;
(c) fixed platforms;
(d) platforms, buckets, boatswain's chairs, etc. suspended from lifting appliances;
(e) safety belts and lifelines; and
(f) catch nets or catch platforms.

29.1.2. Prefabricated parts should be so designed and made that they can be safely transported and erected.

29.1.3. In addition to the conditions of stability of the part when erected, when necessary to prevent danger the design should explicitly take into account:

(a) the conditions and methods of attachment in the operations of stripping, transport, storing and temporary support during erection; and
(b) methods for the provision of safeguards such as railings and working platforms, and, when necessary, for mounting them easily on prefabricated parts.

29.1.4. The hooks and other devices incorporated in prefabricated parts that are required for lifting and transporting them should be so shaped, dimensioned and positioned as:

(a) to withstand with a sufficient margin the stresses to which they are subjected; and
(b) not to set up in the part stresses that could cause failures, or stresses in the building not provided for in the plans.

29.1.5. Prefabricated parts made of concrete should not be stripped before the concrete has set and hardened sufficiently to ensure the safety of the operation.

29.1.6. Storeplaces should be so constructed that:

(a) there is no risk of prefabricated parts falling or overturning; and
(b) storage conditions generally ensure stability having regard to the method of storage and atmospheric conditions.

29.1.7. Prefabricated parts made of concrete should not be erected before the concrete has set and hardened to the extent provided for in the plans.

29.1.8. While they are being stored, transported, raised or set down, prefabricated parts should not be subjected to stresses prejudicial to their stability.

29.1.9. Every lifting appliance should:

(a) be suitable for the operations; and
(b) be approved by a competent person, or tested under a proof load 20 per cent heavier than the heaviest prefabricated part.

29.1.10. Lifting hooks should be of the self-closing type or of a safety type.

29.1.11. Lifting hooks should have the maximum permissible load marked on them.

29.1.12. Tongs, clamps and other appliances for lifting prefabricated parts should:

(a) be of such shape and dimensions as to ensure a secure grip without damaging the part; and
(b) be marked with the maximum permissible load in the most unfavourable lifting conditions.

29.1.13. Prefabricated parts should be lifted by methods or appliances that prevent them from spinning accidentally.

29.1.14. While prefabricated parts are being lifted measures should be taken to prevent workers from being struck by objects falling from a height.

29.1.15. Prefabricated parts should be free from ice and snow at the time of erection.

29.1.16. When necessary to prevent danger, before they are raised from the ground, prefabricated parts should be provided with safety devices such as railings and working platforms to prevent falls of persons.

29.1.17. While prefabricated parts are being erected the workers should be provided with and use appliances for guiding them as they are being lifted and set down, so as to avoid crushing the hands and to facilitate the operations.

29.1.18. Before it is released from the lifting appliance a raised prefabricated part should be so secured that its stability cannot be imperilled, even by external agencies such as wind and passing loads.

29.1.19. If workers are exposed to danger when releasing prefabricated parts from lifting appliances, adequate safety measures should be taken.

29.1.20. At workplaces adequate instruction should be given to the workers on the methods, arrangements and means required for the construction, storage, transport, lifting and erection of prefabricated parts.

Transport

29.1.21. During transport, attachments such as slings and stirrups mounted on prefabricated parts should be securely fastened to the parts.

29.1.22. Prefabricated parts should be so transported that no conditions affecting the stability of the parts or the means of transport result from jolting, vibration, or stresses due to blows, or loads of material or persons.

Placing of prefabricated parts

29.1.23. When the method of erection does not permit the provision of other means of protection against falls of persons, the workplaces should be protected by guard-rails, and if appropriate by toe-boards, in conformity with the provisions of paragraphs 2.6.1 to 2.6.5.

29.1.24. When parts of outside walls are being placed, the area below into which the part might fall should be barricaded or guarded.

29.1.25. When weather conditions such as wind or reduced visibility entail risks of accidents the work should be carried on with particular care, or, if necessary, interrupted.

29.1.26. Before erecting prefabricated parts from work-places from which a person could fall more than 2 m (6 ft 6 in), guard-rails should be provided, completed if necessary by toe-boards.

29.1.27. When it is not practicable to install protective guard-rails and toe-boards the workers should be provided with and use safety belts and lifelines to limit the height of the fall.

29.1.28. The safety devices (guard-rails, toe-boards, safety belts and lifelines) should not be removed so long as the risk remains.

29.2. Structural steel erection

General provisions

29.2.1. As far as practicable the safety of structural steel erectors should be ensured by appropriate means, such as provision and use of:

(a) ladders;

(b) gangways;

(c) fixed platforms;

(d) platforms, buckets, boatswain's chairs, etc. suspended from lifting appliances;

(e) safety belts and lifelines; and

(f) catch nets or catch platforms.

29.2.2. Structural steel erection operations should comply with the relevant provisions of section 29.1 wherever applicable, and in addition the provisions below should be observed whenever they are stricter.

29.2.3. Steel structures should not be worked on during violent storms or high winds, or when they are covered with ice or snow, or are slippery from other causes.

29.2.4. If necessary to prevent danger, structural steel parts should be equipped with attachments for suspended scaffolds, lifelines or safety belts and other means of protection.

29.2.5. If necessary to prevent danger, written instructions concerning the lifting, transport, erection and storage of structural steel parts should be provided and kept available on the site.

29.2.6. Structural steel parts that are to be erected at a great height should as far as practicable be assembled on the ground.

29.2.7. When structural steel parts are being erected, a sufficiently extended area underneath the workplace should be barricaded or guarded.

29.2.8. Suitable devices should be used to raise or lower structural steel parts.

29.2.9. Devices referred to in paragraph 29.2.8 should be used in a manner to prevent slipping of the structural steel part raised or lowered.

29.2.10. Before structural steel parts are lifted, care should be taken that any object that could fall is fastened or removed.

29.2.11. Structural steel parts should not be dragged while being lifted if that could cause danger.

29.2.12. Steel trusses that are being erected should be adequately shored, braced or guyed until they are permanently secured in position.

29.2.13. No load-bearing structural member should be dangerously weakened by cutting, holing or other means.

29.2.14. Workers cleaning steel surfaces with harmful substances should wear appropriate protective equipment complying with relevant requirements of Chapter 36.

Flooring

29.2.15.   If steel erection work is performed from a temporary working floor, workers employed directly underneath should be protected by a close-planked floor above them.

29.2.16.   If steel is stored on temporary flooring, the flooring should be strong enough to support it safely, be close planked, and be adequately secured in position, and the load should be evenly distributed.

Hoisting

29.2.17.   While structural members are being moved into place the load should not be released from the hoisting rope until the members are securely fastened in place.

29.2.18.   Structural members should not be forced into place by the hoisting machine while any worker is in such a position that he could be injured by the operation.

29.2.19.   Open-web steel joists that are hoisted singly should be directly placed in position and secured against dislodgement.

29.2.20.   Bundles of joists should be secured against dislodgement after being hoisted.

29.2.21.   No load should be placed on open-web steel joists until they have been placed in position and secured.

29.2.22.   While panels or structural members are being hoisted, hand ropes should be used to maintain control of the load.

Riveting

29.2.23.   When hot riveting of structural steel is to be done in a confined space the steel surfaces in the area of operations should be free of any compound containing lead or any other substance that could liberate toxic fumes when heated.

29.2.24.   Rivet holes should be cleared of paint by reaming or other effective method.

29.2.25.   Rivets, bolts, nuts, wrenches and other loose articles should be kept in boxes or otherwise prevented from falling from a height.

29.2.26. Precautions should be taken to prevent fires being caused by rivet-heating equipment.

29.2.27. Pails of water or other fire-extinguishing appliances should be provided for rivet heaters.

29.2.28. Rivet heaters should extinguish their fires before leaving work.

29.3. Work in shafts and stair wells

29.3.1. In buildings where stair wells or other shafts are being constructed of masonry, floor slabs or planking should be installed as the work progresses and should be maintained within two floors of the floor on which any brickwork or masonry is being erected.

29.3.2. During the installation or replacement of elevators, or other work in a shaft, a close-planked platform should be installed one floor above the floor at which the work is being done so as to provide protection against falls of objects.

29.3.3. A close-planked platform should be installed at the top of the shaft to prevent persons from falling into it.

29.3.4. Where necessary to prevent danger, an adequate partition should be provided to prevent contact with an adjacent elevator or counterweight when an elevator is being replaced.

29.3.5. When necessary to prevent danger, while men are working at intermediate levels in a shaft:

(a) a catch net should be installed not more than 6 m (20 ft) below the working level; or

(b) the workers should wear safety belts complying with the relevant requirements of Chapter 36.

29.4. Erection of roof trusses

29.4.1. When roof trusses are being erected:

(a) the work should be done from a close-planked floor; or

(b) suitable scaffolding or boatswain's chairs should be provided; or

(c) the workers should be protected by other effective means.

29.4.2. For raising heavy trusses suitable masts or other lifting appliances should be used.

29.4.3. Parts of trusses or columns should be adequately braced or guyed if necessary to prevent danger before the truss is permanently stable.

29.4.4. Workers should not walk along the top members of trusses.

29.4.5. If workers have to walk along the bottom members of trusses a gangway complying with the requirements of section 3.3 should be provided.

29.5. Provision of temporary floors

29.5.1. All tiers of open joists and girders on which workers are employed should be securely covered with close planking or any other effective covering until the permanent floor is installed.

29.5.2. Parts of the protection should only be removed to the extent required for the continuation of the work.

29.5.3. In halls and similar buildings without intermediate walls, columns or chimneys, close planking may be replaced by working platforms complying with the requirements of section 3.2 and at least 2 m (6 ft 6 in) wide.

29.5.4. Adequate precautions should be taken to prevent workers from going on or loading light false floors, vaulting or ceilings.

29.5.5. In buildings or structures of skeleton steel construction, permanent floor filling should as far as practicable be installed as the erection progresses.

29.6. Work on tall chimneys

Scaffolds

29.6.1.   For the erection and repair of tall chimneys, scaffolding complying with the relevant requirements of Chapter 3 should be provided.

29.6.2.   Fixed inside scaffolding should be securely anchored in the chimney wall.

29.6.3.   The scaffold floor should always be at least 65 cm (26 in) below the top of the chimney.

29.6.4.   Under the working floor of the scaffolding the next lower floor should be left in position as a catch platform.

29.6.5.   Suspended outside scaffolding (inspection scaffolds) should comply with the requirements of section 3.5 or section 3.6, as the case may be.

29.6.6.   The distance between the inside edge of the scaffold and the wall of the chimney should not exceed 20 cm (8 in) at any point.

29.6.7.   An adequate catch net should be maintained at a suitable distance below the scaffold.

29.6.8.   Bracket scaffolds should only be used for light work.

29.6.9.   If bracket scaffolds are used the workers using them should be secured by safety belts conforming to the relevant requirements of Chapter 36; the lifeline should be anchored at a point other than that where the brackets are attached.

29.6.10.   When mobile inside suspended scaffolds are used:

(a) the lifting appliance should be firmly secured against displacement;

(b) the working platform with the catch platform underneath should be permanently attached to the lifting appliance; and

(c) when the working platform is not being moved, it should rest on adequate supports securely fastened in the chimney wall.

Catch platforms

29.6.11.   Catch platforms should be erected over:

(a) the entrance to the chimney; and

(b) passageways and working places where workers could be endangered by falling objects.

Stairs, ladders

29.6.12.   For climbing tall chimneys, workers should be provided with:

(a) stairs or ladders complying with the relevant provisions of Chapter 4; or

(b) a column of iron rungs securely embedded in the chimney wall.

29.6.13.   Ladderways should be separated from the hoisting shaft throughout their length by a partition.

29.6.14.   Iron rungs should:

(a) be spaced between 25 cm (10 in) and 30 cm (12 in) apart; and

(b) be at the most 45 cm (18 in) long.

29.6.15.   Iron rungs should be adequately protected against corrosion.

29.6.16.   If the chimney is over 6 m (20 ft) high, workers should use a safety climbing belt to climb the chimney.

29.6.17.   When workers use the outside rungs to climb the chimney, a securely fastened steel core rope looped at the free end and hanging down at least 3 m (10 ft) should be provided at the top to help the workers to climb on to the chimney.

29.6.18.   The climbing of chimneys by means of hooks, trestles, rope ladders, ropes or other climbing aids should be prohibited.

Lifting appliances

29.6.19.   Winches and other lifting appliances should comply with the relevant provisions of this Code, in particular those contained in Chapter 5; and with the requirements of paragraphs 29.6.20 to 29.6.25.

29.6.20. Power-driven lifting appliances should only be used inside chimneys whose clear width at working places is at least 1.5 m (5 ft).

29.6.21. Outriggers for gin wheels and pulley blocks of lifting appliances should be securely anchored in the chimney wall and not be fastened to scaffolding.

29.6.22. Stiff-leg derricks should be protected against splitting and slipping.

29.6.23. On the working platform the opening for the lifting appliance should be protected by guard-rails and toe-boards complying with the requirements of paragraphs 2.6.1 to 2.6.5.

29.6.24. Sections of the guard-rails and toe-boards that are removed for attaching and detaching the bucket should be secured in position when the bucket is not being handled.

29.6.25. Workers should not be carried on lifting appliances intended only for materials.

Personal protective equipment

29.6.26. Workers employed on independent chimneys should wear safety helmets and, when necessary to prevent danger, a safety belt complying with the relevant requirements of Chapter 36.

Signalling

29.6.27. When necessary to prevent danger there should be adequate means of signalling or other means of communication between the working levels and the ground.

Operations

29.6.28. While work is being done on independent chimneys the area surrounding the chimney should be enclosed by fencing at a safe distance.

29.6.29. No unauthorised person should enter the area referred to in paragraphs 29.6.28.

29.6.30. Workers employed on the construction, alteration, maintenance or repair of tall chimneys should not:

(a) work on the outside without a safety belt attached by a lifeline to a rung, ring or other secure anchorage;

(b) put tools between the safety belt and the body or in pockets not intended for the purpose;

(c) haul heavy materials or equipment up and down by hand to or from the workplace on the chimney;

(d) fasten pulleys or scaffolding to reinforcing rings without first verifying their stability;

(e) work alone;

(f) climb a chimney that is not provided with securely anchored ladders or rungs; or

(g) work on chimneys in use unless the necessary precautions to avoid danger from smoke and gases have been taken.

29.6.31.   Materials, tools, etc. should be hoisted in containers from which they cannot fall.

29.6.32.   Bricks should not be hoisted in rope slings.

29.6.33.   Care should be taken that loads being raised do not catch on any fixed object.

29.6.34.   Work on independent chimneys should not be carried on in high winds, in icy conditions, in fog or during electrical storms.

29.7. Work with hot asphalt, tar, etc.

Equipment

29.7.1.   Asphalt plants should comply with the requirements of section 15.5.

29.7.2.   Tanks, vats, kettles, pots and other vessels for heating tar, pitch, asphalt and other bituminous materials should:

(a) be adequately resistant to damage by heat;

(b) be capable of holding a full load without danger of collapse, bursting, distortion or tipping; and

(c) be provided with a close-fitting cover suitable for smothering a fire in the vessel.

29.7.3. Gas and oil-fired asphalt and tar kettles or pots should be equipped with burners, regulators and adequate safety devices.

29.7.4. Heating appliances for vessels should distribute the heat uniformly over the heating surface so as to avoid overheating.

29.7.5. Only electrically heated vessels should be used inside buildings; but if other types of heating are employed, special safety precautions should be observed.

29.7.6. Buckets for hot asphalt, bituminous materials or tar should have:

(a) the bail or handle firmly secured; and
(b) a second handle near the bottom for tipping.

29.7.7. When vessels containing asphalt solutions are heated by gas burners or other fuel producing open flames, the flame should be of the jet or pressure burner type and be enclosed by a hood or surrounded by baffles.

29.7.8. While vessels containing asphalt solutions are being heated they should be kept open.

Operations

29.7.9. Asphalt should be kept dry.

29.7.10. Workers employed at vessels or handling hot bituminous materials should be provided with suitable protective equipment.

29.7.11. Vessels in operation should be kept at a safe distance from combustible materials.

29.7.12. When vessels are used in confined spaces the gases, fumes and smoke generated should be removed:

(a) by exhaust ventilation; or
(b) if exhaust ventilation is not adequate, by forced ventilation.

29.7.13. Vessels that are being heated should not be left unattended.

29.7.14.   Pieces of asphalt or other bituminous material should not be thrown into hot vessels so as to cause dangerous splashing.

29.7.15.   Covers should be kept closed when vessels are not in use.

29.7.16.   Buckets should not be filled with hot asphalt or other bituminous materials to a level that might cause danger when they are carried or hoisted.

29.7.17.   Workers should not carry buckets of hot bituminous material on ladders.

29.7.18.   Mops and other applicators contaminated with bituminous materials should not be stored inside buildings.

29.8. Work with wood preservatives

29.8.1.   Before any work is done with wood preservatives the employer should ascertain whether the preservatives contain any chlorinated derivatives of phenol or naphthol or salts of arsenic and chromium or other dangerous substances.

29.8.2.   If wood preservatives contain dangerous substances the requirements of paragraphs 29.8.3 to 29.8.9 should be complied with.

29.8.3.   As far as practicable, wood should be impregnated with preservatives before it reaches the construction site.

29.8.4.   Wood that has been impregnated by dipping on the construction site or in the timber yard should not be used before it has dried sufficiently or, in special cases, the precautions required by paragraphs 29.8.5 to 29.8.7 have been taken.

29.8.5.   When work is being done with wood preservatives, including creosote, the necessary personal protective equipment should be provided and used as follows:

(a) for dipping: protective clothing, gauntlets and eye protection and, if necessary, protection of exposed skin;

(b) for brush painting: gauntlets, and, if splashing may occur, eye protection and, if necessary, protection of exposed skin;

(c) for spraying: protective clothing, respiratory protection and eye protection and, if necessary, protection of exposed skin.

29.8.6.   Workers using wood preservatives should be provided with suitable barrier creams not soluble in oil, or other appropriate creams, for applying to the exposed parts of the skin.

29.8.7.   The creams referred to in paragraph 29.8.6 should be applied before the work begins.

29.8.8.   Creosote should not be allowed to come into contact with the skin or eyes.

29.8.9.   Creosote burns should be washed immediately and receive medical attention.

29.9. Laying floors, facing walls, etc. with flammable materials

29.9.1.   When in work such as laying or treating floors, gluing sheets of linoleum, plastic, rubber, etc. on floors, walls or equipment, use is made of glue or other highly flammable material, the requirements of this section should be complied with during the work and the drying time.

29.9.2.   The place or space where the work is being done should be well ventilated, for example, by a through draught or a suitable fan.

29.9.3.   When a fan is used it should be placed outside the place or space.

29.9.4.   Before the work begins, outside the entrance to the place or space a suitable warning notice should be affixed.

29.9.5.   Smoking, gas and electric welding, and cutting with gas or other open flame should be prohibited in the place or space and its vicinity.

29.9.6.   Doors to adjoining places or spaces where there are open flames should be kept closed.

29.9.7.   All inside electrical installations should be flameproof.

29.9.8.   Fire-extinguishing equipment should be provided in accordance with the relevant provisions of this Code.

29.10. Insulation work

Work with asbestos

29.10.1. Every effort should be made to replace materials containing asbestos by less dangerous materials.

29.10.2. When material containing asbestos is used the material should be kept wet.

29.10.3. When material containing asbestos is used:

(a) equipment processing or handling the material should be enclosed; or

(b) unenclosed equipment should be provided with exhaust ventilation; or

(c) the workers should wear respiratory protective equipment.

29.10.4. When asbestos compounded slabs are being processed in a stationary machine the machine should be equipped with a system for exhausting the dust.

29.10.5. When asbestos compound slabs are being processed by a portable machine or by hand, the workers should be protected, if necessary to prevent danger, by exhaust ventilation or respiratory protective equipment.

29.10.6. When insulating blocks, slabs, cloth, tapes, cords, etc. containing asbestos are used, the workers should wear respiratory protective equipment, and cloth, tapes, cords and the like should be kept wet.

29.10.7. Insulating mattresses enclosed and filled with asbestos should be prepared in enclosed equipment provided with exhaust ventilation.

29.10.8. When asbestos is sprayed in buildings:

(a) machines for shredding and spreading asbestos should be equipped with means of keeping the asbestos wet;

(b) valves for the water and asbestos supplies should be so interlocked in spraying equipment that the water must be turned on first and the asbestos turned off first.

29.10.9. When asbestos is sprayed in buildings:

(a) the workers should wear respiratory protective equipment;

(b) all workers not employed on spraying should be at a safe distance.

29.10.10.   When asbestos is sprayed in buildings, asbestos dust deposited on the floor, ledges, etc. should be kept wet and removed as soon as practicable, and in any case at the end of the shift.

Work with glass wool and similar materials

29.10.11.   When insulating work is being done with glass wool, glass fibres, mineral wool and the like, the workers should wear protective clothing that, as far as practicable, prevents such substances from coming into contact with the skin.

29.10.12.   Protective clothing worn during work with insulating material, such as glass fibres, mineral wool and the like, should not be utilised as street clothes.

29.11. Window cleaning

General provisions

29.11.1.   Safety devices such as anchors, safety belts, scaffolds, boatswain's chairs and ladders should be provided for window cleaners at all windows through which they could fall more than 2 m (6 ft 6 in).

29.11.2.   Ladders should not be used for window cleaning if the windows can be cleaned safely by other means.

29.11.3.   No person should clean a window having broken sash chains or cords.

29.11.4.   No person should clean a window for which anchors are required unless he is wearing a safety belt complying with the relevant requirements of Chapter 36 and has effectively secured it to the anchors.

29.11.5.   If a window for which anchors are required has a sill that is not safe for the cleaner to stand on, a safe auxiliary sill or other device enabling the cleaner to stand safely should be provided.

29.11.6. Anchors should not be deemed to be adequate safety devices unless the window and its approaches are so constructed that the cleaner can reach the sill and attach one terminal of the belt to the anchor before stepping on the sill, and one terminal can remain attached to the anchor while the cleaner steps back into the building.

29.11.7. No corrosive substance other than dilute hydrochloric acid or dilute hydrofluoric acid should be used for window cleaning.

29.11.8. If acid is used for window cleaning:

(a) scaffolding complying with the requirements of Chapter 3 should be used;
(b) no fibre rope should be used on the scaffold;
(c) the window cleaner should wear rubber gloves;
(d) the acid should be applied with a brush; and
(e) all parts of the scaffold and other equipment used should be thoroughly washed with water after each use.

29.11.9. Where necessary to prevent danger to window cleaners, windows more than 1.80 m (6 ft) wide should be provided with a safe back rest, such as a cable.

29.11.10. Scaffolds and boatswain's chairs for window cleaners should comply with the relevant requirements of Chapter 3.

29.11.11. Ladders for window cleaners should comply with the relevant requirements of Chapter 4.

29.11.12. A window cleaner should not pass from one window to another on the outside of a building unless he is on a safe support all the time and one end of his safety belt is continuously attached.

Anchors

29.11.13. Anchors should be forged or machined from corrosion-resistant rolled metal alloy.

29.11.14. The forging should be given suitable heat treatment.

29.11.15.   Anchors should be provided with means that prevent them from inadvertently turning, backing off, or becoming loose.

29.11.16.   Windows more than 1.80 m (6 ft) wide for which an anchor is required should have two anchors at each end for fastening the window cleaner's safety belt.

29.11.17.   Before being taken into use, anchors should be tested by a competent person and found to be safe for use.

29.11.18.   No window cleaner should work on a window that has a loose or otherwise insecure or missing anchor.

29.12. Stone dressing

29.12.1.   Structures, machines and appliances used in stone dressing should be so designed or protected that, as far as practicable, workers using them are not exposed to danger from siliceous dust.

29.12.2.   Enclosed workplaces for stone-dressing operations with a silicosis risk should:

(a) have an adequate air space;
(b) be adequately ventilated by dust-free and clean air;
(c) be equipped with adequate dust-collection or dust-suppression appliances; and
(d) be cleaned daily by wet or exhaust methods.

29.12.3.   Workplaces in the open air or in open-sided sheds for stone-dressing operations with a silicosis risk should:

(a) be separated one from another by a distance of at least 4 m (13 ft); and
(b) be so placed with regard to the prevailing wind direction that neighbouring workplaces are not endangered by siliceous dust.

29.12.4.   If workers cannot be protected from siliceous dust by other means they should be provided with suitable respirators.

29.12.5.   Workers dressing stone should take adequate precautions to prevent:

(a) persons in the vicinity from being injured by flying stone splinters;

(b) injury to themselves from splinters from neighbouring workplaces; and

(c) eye injury.

29.12.6. If necessary to prevent danger, stone dressers' workplaces should be separated by stout partitions or screens.

30. Demolition

30.1. Preparatory work

30.1.1. Before demolition operations begin:
(a) adequate inspection should be made; and
(b) if necessary to prevent danger, unstable parts of the building
should be made secure.

30.1.2. Before demolition operations begin, the power on all
electric service lines should be shut off and the lines cut or disconnected at or outside the property line.

30.1.3. Before demolition operations begin, all gas, water
and steam service lines should be shut off and capped or otherwise
controlled at or outside the building line.

30.1.4. If it is necessary to maintain any electric power, water
or other service lines during demolition operations, they should be
adequately protected against damage.

30.1.5. The danger zone round the building should be adequately fenced off or signposted.

30.2. General provisions on demolition operations

30.2.1. All demolition operations should be under the supervision of a competent person.

30.2.2. Demolition operations should only be carried out by
competent workers.

30.2.3. Demolition operations should begin by the removal of:
(a) glass in doors, windows, etc.;
(b) loose objects; and
(c) projecting parts.

30.2.4. Workers should not be employed at different levels
unless adequate precautions are taken to ensure the safety of those
at lower levels.

30.2.5. Demolition operations should begin at the top of the structure and proceed downwards.

30.2.6. Masonry and other material should not be allowed to accumulate or fall in quantities liable to endanger the stability of any floor or structural support.

30.2.7. Care should be taken not to demolish any parts which would destroy the stability of other parts.

30.2.8. Demolition activities should not be continued under climatic conditions, such as high winds, which could cause collapse of already weakened structures.

30.2.9. When necessary to prevent danger, parts of structures should be adequately shored, braced or otherwise supported.

30.2.10. Structures should not be left in a condition in which they could be brought down by wind pressure or vibration.

30.2.11. Where necessary to keep down dust, buildings being demolished should be sprayed with water at suitable intervals.

30.2.12. If parts of buildings are pulled down:

(a) only adequate wire cables should be used; and
(b) all workers should be at a safe distance.

30.2.13. Special precautions should be taken when any structure is being demolished by undermining.

30.2.14. If buildings or parts of buildings are demolished by explosives, the relevant requirements of Chapters 22 and 23 should be complied with.

30.2.15. Foundation walls serving as retaining walls to support earth or adjoining structures should not be demolished until the adjoining structure has been underpinned or braced, and the earth removed or supported by sheet piling or sheathing.

Areas of access and egress

30.2.16. Safe passageways should be provided for workers on demolition sites.

30.2.17. Stairs should be kept in place as long as practicable.

30.2.18.   So long as stairs are in place, stair railings should not be removed.

30.2.19.   When necessary to prevent danger, all passageways, stairs, ladderways and other parts of a building where workers have to work or pass should be adequately lit.

Personal protective equipment

30.2.20.   Workers employed in demolition operations should wear safety boots, safety helmets and heavy gloves.

30.2.21.   Workers employed in dusty operations should wear respirators of an appropriate type.

30.3. Demolition equipment

30.3.1.   When equipment such as power shovels and bull-dozers is used for demolition, due consideration should be given to the nature of the building or structure, its dimensions, etc., as well as to the power of the equipment being used.

30.3.2.   If a swinging weight is used for demolition, a safety zone having a width at least 1½ times the height of the building or structure should be maintained around the points of impact.

30.3.3.   Swinging weights should be so controlled that they cannot swing against any structure other than the one being demolished.

30.3.4.   If a clamshell bucket is used for demolition, a safety zone extending 8 m (25 ft) from the line of travel of the bucket should be maintained.

30.3.5.   No person other than a worker operating the demolition equipment should enter the safety zone.

30.3.6.   Mechanical demolition equipment should be operated from a safe place.

30.3.7.   If derricks are used for the demolition of structural steelwork, they should be installed and operated in conformity with the requirements of section 5.8.

30.3.8. Scaffolds used for demolition operations should, to ensure their stability, be independent of the structure to be demolished.

30.3.9. If ladders are required, only travelling mechanical ladders should be used, except that ordinary ladders may be used as means of access to the scaffolds referred to in paragraph 30.3.8.

30.3.10. Ladders should not lean against any part of the structure to be demolished.

30.3.11. Wherever practicable, hoists or chutes should be provided to lower material.

30.3.12. Material chutes should have a gate at the bottom with suitable means for regulating the flow of material.

30.4. Catch platforms for demolition operations

30.4.1. Where necessary during the demolition of buildings or other structures, catch platforms should be provided along the outside of exterior walls so as to prevent danger from falling objects; provided that they may be dispensed with if a sufficient safety zone is established.

30.4.2. Where necessary, catch platforms should be moved as demolition proceeds so as to be always not more than two floors below the demolition level until demolition is within three floors of ground level.

30.4.3. Catch platforms should be at least 1.5 m (5 ft) wide and close planked.

30.4.4. Catch platforms should be inclined so that the outer edge is higher than the inner edge.

30.4.5. Catch platforms should be able to withstand safely a live load of 600 kg/m² (120 lb/ft²).

30.5. Demolition of walls

30.5.1. Walls should be demolished storey by storey beginning at the roof and working downwards.

30.5.2. Masonry and brickwork should be taken down in reasonably even courses.

30.5.3. Where necessary, unsupported walls should be protected against falling by means such as shoring and ties.

30.5.4. If walls are pushed over, workers should be kept at a safe distance and, if necessary to prevent danger, be protected against flying fragments.

30.5.5. Workers demolishing walls from which they could fall a dangerous distance should be provided with scaffolding, or protected by a catch platform or other effective means.

30.5.6. Workers demolishing walls that are thin or structurally weak should be provided with scaffolding.

30.5.7. Walls should not be subjected to dangerous lateral pressure by stored material.

30.6. Demolition of floors

30.6.1. When necessary to prevent danger, workers demolishing floors should be provided with planking or walkways on which to stand or move.

30.6.2. Openings through which material is dropped should be adequately fenced or barricaded to prevent danger.

30.6.3. Floor openings for ladders or stairs should be provided with guard-rails and toe-boards complying with the requirements of paragraphs 2.6.1 to 2.6.5.

30.6.4. When floors are being demolished, the area immediately underneath should be fenced off, and no worker should be allowed to enter it.

30.6.5. All work above each tier of floor beams should be completed before the safety of the tier supports is impaired.

30.7. Demolition of structural steelwork

30.7.1.   All practicable precautions should be taken to prevent danger from any sudden twist, spring or collapse of steelwork, ironwork or reinforced concrete when it is cut or released.

30.7.2.   Steel construction should be demolished tier by tier.

30.7.3.   Structural steel parts should be lowered and not dropped from a height.

30.8. Demolition of tall chimneys, steeples, etc.

30.8.1.   Tall chimneys should not be demolished by blasting or overturning unless a protected area of adequate dimensions can be established in which the chimney can fall safely.

30.8.2.   Tall chimneys should only be demolished by competent persons under constant competent supervision.

30.8.3.   If tall chimneys are demolished by hand, scaffolding complying with the relevant requirements of Chapter 3 should be used.

30.8.4.   The scaffolding should be shifted as demolition proceeds so that the working platform is always more than 25 cm (10 in) but not more than 1.5 m (5 ft) below the top of the chimney.

30.8.5.   Supports for hoisting appliances should be independent of the scaffolding.

30.8.6.   Workers should not stand on top of the chimney wall.

30.8.7.   If material is thrown down inside the chimney, an appropriate opening should be made at the bottom to prevent accumulations.

30.8.8.   Material thrown down should only be removed during breaks in the work.

30.8.9.   If workers are hoisted for the demolition of chimneys, they should be conveyed:

(a) in a boatswain's chair complying with the requirements of section 3.17; or

(b) in an equally safe manner.

30.8.10. The relevant provisions of this section should apply also to the demolition of steeples and similar structures.

31. Excavations

31.1. General provisions

31.1.1. Before excavation begins on any site, the stability of the ground should be verified by a competent person.

31.1.2. Before work begins on any excavation site, the employer should verify the position of all underground installations such as sewers, gas pipes, water pipes and electrical conductors that may cause danger during the work.

31.1.3. If necessary to prevent danger before an excavation is started, gas, water, electrical and other public utilities should be shut off or disconnected.

31.1.4. If underground pipes, conductors, etc. cannot be removed or disconnected, they should be fenced, hung up or otherwise protected.

31.1.5. If necessary to prevent danger, land should be cleared of trees, boulders and other obstructions before excavation begins.

31.1.6. Sides of excavations should be thoroughly inspected:

(a) after an interruption in work of more than one day;
(b) after every blasting operation;
(c) after an unexpected fall of ground;
(d) after substantial damage to supports;
(e) after a heavy frost;
(f) after heavy rain; and
(g) when boulder formations are encountered.

31.1.7. Safe means of access and egress should be provided to every place where persons are employed in excavations.

31.1.8. No person should work on loose ground if the slope is too steep to ensure a safe foothold.

31.1.9. When ground does not ensure a safe foothold, adequate support should be provided.

31.1.10. No ground should be undermined without adequate support.

31.1.11. No person should work under an overhanging or undermined tree stump, wall or other structure.

31.1.12. When loose masses or large boulders and rocks are encountered:

(a) they should be removed as soon as practicable from above; and

(b) workers should leave and keep out of the danger zone until it is safe to return.

31.1.13. Where persons are working at different levels, adequate means, such as flooring, should be provided to prevent persons below from being struck by tools or other objects falling from above.

31.1.14. Openings in flooring referred to in paragraph 31.1.13 should be provided with covers opening upwards and kept closed when not in use.

31.1.15. If necessary to prevent danger, sides of excavations and piles of excavated material should be adequately lit during the hours of darkness.

31.1.16. As far as practicable, excavations should be kept free from water.

31.1.17. In any excavation where there is reason to fear danger from inrushes of water or falls of material, a safe way of escape should, as far as practicable, be provided for every worker.

31.1.18. No person should enter a sewer, shaft or other underground space or chamber unless it has been tested and found free from dangerous quantities of harmful gases.

31.1.19. If persons have to enter an underground chamber or other place to test for gas, they should be provided with a safety belt, safety line and breathing apparatus complying with the relevant requirements of this Code.

31.1.20. If necessary to prevent danger, adequate mechanical ventilation should be provided in excavations to disperse harmful gases and fumes.

31.1.21. When internal combustion engines are operated in an excavation, steps should be taken to avoid the accumulation of dangerous gases by providing exhaust scrubbers, improved ventilation or other effective means.

31.1.22. Every accessible part of an excavation into which a person could fall should, where necessary, be protected by an adequate barrier.

31.1.23. No material should be placed or stacked near the edge of any excavation so as to endanger persons employed below.

31.1.24. No load, plant or equipment should be placed or moved near the edge of any excavation where it would be likely to cause a collapse of the side of the excavation and thereby endanger any person.

31.1.25. If an excavation is likely to affect the security of a structure on which persons are working, precautions should be taken to protect the structure from collapse.

31.2. Support of excavation workings

31.2.1. Sides of excavations where workers are exposed to danger from moving ground should be made safe by sloping, shoring, portable shields or other effective means.

31.2.2. An adequate supply of timber or other suitable shoring material should be available where excavation work is being carried out.

31.2.3. Timber and other supports should only be erected, substantially altered or dismantled under the supervision of a competent person and by competent workers.

31.2.4. All struts, braces and wallings in excavations should be adequately secured so as to prevent their accidental displacement.

31.2.5. If necessary to prevent danger, masonry walls protecting excavations should be adequately braced or shored.

31.2.6. Temporary sheet piling installed for the construction of a retaining wall should not be removed until the wall has attained its full strength.

31.2.7. Earth banks should not be undercut unless they are adequately shored.

31.2.8. Heavy equipment such as power shovels and derricks should not be placed near the edges of excavations unless precautions such as the provision of shoring or piling are taken to prevent the sides from collapsing.

31.2.9. In frozen ground, shoring should not be dispensed with unless the depth and duration of frost and the consistency of the soil are sufficient to prevent danger of collapse.

31.3. Trenches

31.3.1. Trenches in built-up areas and on traffic routes should be fenced.

31.3.2. Depending on the type of soil, the sides of trenches should be secured against falling in by adequate sloping, shoring, portable shields or other effective means.

31.3.3. If necessary to prevent danger, workers installing shoring should be protected by frames, bracing or other effective means.

31.3.4. Trenches more than 1.20 m (4 ft) deep should be provided with ladders at suitable intervals.

31.3.5. Ladders should extend from the bottom of the trench to at least 90 cm (3 ft) above the ground.

31.3.6. Where workers use hand tools such as picks and shovels they should keep a safe distance apart from each other while working in trenches.

31.3.7. When a mechanical digger is used for trenching, the timbering should follow the digger as closely as practicable.

31.3.8.   Trenches in unstable ground such as loose sand should be close timbered.

31.3.9.   Foot-boards and platforms supported by bracing should be adequately secured by brackets or other effective means.

31.3.10.   Bracing should not be used as a ladder.

31.3.11.   Heavy objects should not be placed on bracing.

31.3.12.   When buckets of hot material are lowered, special precautions should be taken to prevent injury to the workers.

31.3.13.   If heating appliances are used in or near trenches in frozen ground, precautions should be taken to ensure that they do not dangerously weaken the sides.

31.3.14.   When trenches are being filled in, the shoring should be kept in place so long as is necessary to prevent danger from collapse of the sides.

31.4. Wells

31.4.1.   Hoisting equipment installed over wells should:

(a) possess adequate strength and stability; and
(b) not endanger workers below.

31.4.2.   Wells should be securely cased or lined to within 1.5 m (5 ft) from the bottom as they are sunk.

31.4.3.   A ladderway complying with the relevant requirements of Chapter 4 should be installed from the top to the bottom of wells as they are sunk.

31.4.4.   In water-bearing ground, wells should be provided with means for the rapid evacuation of the workers.

31.4.5.   If a well has to be continuously pumped out, a reserve set of pumping equipment should be kept available.

31.4.6.   Buckets of earth should be guided while being hoisted, if necessary to prevent danger.

31.4.7.   Workers should not remain at the bottom of a well in which a grab is working.

32. Underground construction

32.1. General provisions

32.1.1.   Tunnelling operations should be carried on in accordance with plans approved by the competent authority.

32.1.2.   All occupied workplaces underground should be inspected at least once in every shift.

32.1.3.   Places occupied by solitary workers should be inspected at least twice in every shift.

32.1.4.   At least once in every week, thorough inspections should be made of all machinery, equipment, structures, supports, roadways, means of egress, magazines, medical facilities, sanitation and working places.

32.1.5.   All workers should be withdrawn from underground workings if:

(a) the ventilation fails; or

(b) other imminent danger threatens.

32.1.6.   Parts of underground workings found to be dangerous should be fenced off.

32.1.7.   A telephone system should be maintained from the vicinity of the face of underground workings to the surface with stations at intermediate workplaces.

32.1.8.   In underground workings that are wet, the workers should be provided with waterproof clothing and boots.

32.1.9.   In tunnels and other underground workings where an explosive mixture of gas such as methane and air may form, operations should be carried on in accordance with national or other official regulations applying to gassy mines.

32.1.10.   In these cases referred to in paragraph 32.1.9, in particular:

(a) all electrical equipment and conductors should be flameproof or intrinsically safe;

(*b*) the current should be cut off from all electrical equipment and conductors when the content of flammable gas in the general body of the air exceeds safe limits;

(*c*) no blasting should be done at any place where the content of flammable gas in the atmosphere exceeds safe limits;

(*d*) suitable gas detectors and gas alarms should be provided;

(*e*) the atmosphere should be tested at suitable intervals in every shift;

(*f*) all workers should be withdrawn from the underground workings when the content of flammable gas in the general body of the air exceeds a percentage to be determined by the competent authority; and

(*g*) no worker should be allowed to carry matches or cigarette lighters underground.

32.2. Shaft sinking

General provisions

32.2.1.   Every shaft not sunk through solid rock should be cased, lined or otherwise made safe.

32.2.2.   Shuttering for masonry lining of shafts should only be removed gradually as the masonry progresses.

32.2.3.   As far as practicable, workers employed in sinking shafts should be protected against falls of objects.

32.2.4.   Workers employed on sinking shafts should be provided with staging, scaffolds or cradles from which they can work safely.

32.2.5.   Staging, scaffolds and cradles should, if necessary for the maintenance of adequate ventilation in the shaft, be provided with grids or other suitable devices.

32.2.6.   As soon as practicable, the shaft top should be protected by adequate fencing, or guard-rails and toe-boards and gates.

32.2.7.   When a shaft is being sunk through water-bearing strata, adequate means of escape from the bottom should be provided.

32.2.8. All entrances between the bottom of a shaft and the top should be securely fenced.

32.2.9. Where a fence or cover has been removed from a shaft entrance to allow work to proceed, it should be replaceable by two horizontal bars, ropes or chains at heights of about 60 cm (2 ft) and 1.20 m (4 ft) from the floor.

32.2.10. All shafts should have a ladderway from the surface to the workings, in addition to any mechanical means of ingress and egress.

32.2.11. Ladders should comply with the relevant requirements of Chapter 4.

32.2.12. Shafts used for hoisting should have a ladder compartment separated from the haulage compartment by fencing adequate to prevent danger.

32.2.13. In blasting operations, all shots should be fired by electricity.

32.2.14. If shaft sinking is carried on at night, the shaft top should be adequately lit.

32.2.15. A thorough inspection of the shaft should be made:
(a) before a shift descends; and
(b) after blasting.

32.2.16. While persons are in the shaft, the bottom should be adequately lit.

Hoisting during shaft sinking

32.2.17. Hoisting installations should comply with the relevant requirements of this Code, in particular, those contained in Chapter 5; and the requirements of paragraphs 32.2.18 to 32.2.31.

32.2.18. An adequate clear space should be provided between the hoisting pulley and the bucket when it is at the top of the shaft.

32.2.19. As soon as practicable, guides for the bucket should be provided.

32.2.20. The bucket should be so fastened to the hoisting rope that it cannot become detached.

32.2.21.   Winches should:

(a) comply with the requirements of section 5.11; and
(b) be provided with an adequate depth indicator.

32.2.22.   Winches at shaft tops should be so installed that the bucket can be attached and detached safely.

32.2.23.   Shafts equipped with a hand-operated winch should have the top protected by a toe-board.

32.2.24.   When persons are raised and lowered in a bucket, at the top and working level the shaft should be closed by doors or flaps, which should only be opened to allow the passage of the bucket or material.

32.2.25.   Hoisting operations in shaft sinking should be governed by signals in conformity with the relevant requirements of section 1.8.

32.2.26.   No person should be hoisted without a light.

32.2.27.   No person should enter or leave the bucket at the top of the shaft or at any working level before the flaps or trap doors at the top or the level in question have been closed.

32.2.28.   No person should be carried in a bucket in which material is carried.

32.2.29.   If two buckets are used, men and materials should not be hoisted at the same time.

32.2.30.   Buckets should not be filled to the brim.

32.2.31.   Objects projecting outside the bucket should be securely fastened to the suspension gear or the hoisting rope.

32.3. Hoisting shaft installations

Shaft tops

32.3.1.   Tops of shafts should be adequately protected against inrushes of water.

32.3.2.   Tops of shafts should be protected by guard-rails and toe-boards complying with the requirements of paragraphs 2.6.1 to 2.6.5.

Shaft supports

32.3.3. Structures in shafts should be:

(a) kept free of stones and other objects; and
(b) cleaned at suitable intervals.

Headframes

32.3.4. All shafts over 30 m (100 ft) in depth should have an adequate headframe.

32.3.5. Headframes should be strong enough to withstand safely the maximum loads that they will have to carry.

32.3.6. Headframes should preferably be of open steelwork construction.

32.3.7. If headframes are of timber, they should be treated to make them fire-resistant.

32.3.8. Headframes should be earthed or otherwise adequately protected against lightning.

Landings

3 2.3.9. All landings in shafts should be provided with gates that effectively close the opening to a height of at least 2 m (6 ft 6 in).

32.3.10. At all landings where it is necessary to cross the shaft, a safe passageway should be provided.

Ladderways

32.3.11. Ladderways in shafts should comply with the relevant requirements of Chapter 4.

32.3.12. Ladderways should:

(a) be adequately lit from top to bottom;
(b) be installed in a separate shaft; or
(c) be installed in a separate compartment of the hoisting shaft; or
(d) not be used while hoisting is being carried on.

Signalling installations

32.3.13. Shafts should be equipped with a signalling system that warns the hoisting engineer when a conveyance passes beyond the safe limit of travel.

32.3.14. It should be possible to exchange signals effectively between all the landings in the shaft.

32.3.15. Before tunnelling operations are begun from a shaft, two separate signalling systems of different types should be installed.

32.3.16. The signal code should be posted in the hoisting machine room and at each landing.

Hoisting machines

32.3.17. Hoisting machines should be equipped:

(a) with an adequate brake that will automatically stop and hold the conveyance if the hoisting power fails; and

(b) with a reliable depth indicator.

32.3.18. Hoisting drums or ropes should be provided with such distance markings as are required for safe operation.

32.3.19. All parts of hoisting machines and other hoisting equipment should be readily accessible for inspection.

32.3.20. All hoisting machines should be inspected at least once a day by the hoisting engineer.

Cages and buckets

32.3.21. Shafts exceeding 30 m (100 ft) in depth should have an installation for conveying persons.

32.3.22. In completed shafts, the installation for conveying persons should be a cage or car.

32.3.23. Cages or cars for conveying persons should be equipped with safety gear that automatically holds the cage or car when fully loaded if the suspension rope breaks or becomes slack.

32.3.24. Cages or cars for conveying persons should:

(a) be solidly enclosed from floor to roof on at least two sides;

(b) be provided on open sides with a gate or other adequate barrier; and

(c) be provided with a roof affording adequate protection against falling objects.

30.3.25. The cage or car roof (or roofs in the case of multiple deck cages) should have a trap door or other suitable emergency exit.

32.3.26. There should be adequate means of blocking the cage or car at every landing.

32.3.27. Buckets used for conveying persons in shafts should:

(a) have no projections on the outside that could catch in an obstruction;

(b) be not less than 1 m (3 ft 3 in) deep;

(c) be provided with adequate means to prevent them from inadvertently tipping; and

(d) not be self-opening.

32.3.28. Unguided or free-hanging buckets should not be raised or lowered at a speed exceeding 30 m (100 ft) a minute.

Wire ropes, chains and accessories

32.3.29. Only steel wire ropes should be used in shaft installations for hoisting operations.

32.3.30. Steel wire ropes should have a safety factor of at least 10.

32.3.31. Only steel wire ropes fitting the hoisting machine drum should be used.

32.3.32. No spliced steel wire rope should be used for hoisting.

32.3.33. All hoisting steel wire ropes should be suitably lubricated to prevent corrosion and wear.

32.3.34. All hooks used with hoisting equipment should be provided with devices to prevent them from becoming detached from the load while in use.

32.4. Hoisting operations

32.4.1.   Notices should be posted at conspicuous places at the hoisting installation stating:

(a) the maximum speed for transporting persons in the shaft; and
(b) the maximum number of persons and the maximum weight of material that may be safely carried in each conveyance.

32.4.2.   Hoisting operations in shafts should be governed by signals in conformity with the relevant requirements of section 1.8.

32.4.3.   Regular hoisting of persons in shafts should be subject to the authorisation of the competent authority.

32.4.4.   When workers are being hoisted at the beginning or end of a shift:

(a) no material should be hoisted in any compartment of the shaft;
(b) a competent person should be in attendance at all shaft landings in use to:

   (i) give the necessary signals to the hoisting engineer;
   (ii) prevent overloading of the conveyance; and
   (iii) observe all required safety precautions; and

(c) the hoisting engineer should be accompanied by another competent person who could operate the hoisting machine in an emergency.

32.4.5.   Cages or cars should be blocked while persons are entering or leaving or material is being loaded or unloaded at landings.

32.4.6.   No person should ride in a conveyance when material or heavy equipment is being conveyed.

32.4.7.   Landing gates should be kept closed except when the conveyance is stationary at the landing.

32.4.8.   In the hoisting machine room, there should be a light or other signal showing whether all the landing gates are closed or not.

32.4.9.   Cages or cars should not be raised or lowered:

(a) at a speed in excess of the authorised maximum; or

(b) with a greater number of persons or a heavier load of material than the authorised maximum.

32.4.10.   After every stoppage of hoisting for repairs or other reason, the conveyance should be run empty up and down the working part of the shaft at least once.

32.4.11.   Except for repair purposes, hoisting operations should not be carried on in shafts any part of which is being repaired.

32.5. Supports

32.5.1.   When necessary to prevent danger, roofs and sides of tunnels and other underground workings should be adequately supported by timbering or other effective means.

32.5.2.   If supports are required, adequate supplies of suitable support material should be kept readily available.

32.5.3.   Supports should be maintained as close as practicable to the tunnel face.

32.5.4.   The sides and roof and supports of tunnels should be inspected at least once in every shift.

32.5.5.   When a tunnel is to be lined with masonry or concrete, the supports should not be removed from any part until it is safe to remove them.

32.5.6.   When supports are being withdrawn or changed, adequate precautions should be taken to prevent danger from loose masses.

32.5.7.   Additional supports should be installed:

(a) when any part of existing supports is found to be deformed; and

(b) when any part of existing supports is being changed.

32.6. Ventilation

32.6.1.   All underground workings should be traversed by a regular air current to keep them in a fit state for working and, in particular:

(a) to avoid excessive rises in temperature;

(b) to dilute harmful dusts, gases and fumes to safe concentrations; and

(c) to prevent the oxygen content of the atmosphere from falling below 17 per cent.

32.6.2.   In all underground workings it should be possible to reverse the air flow.

32.6.3.   When natural ventilation is inadequate, mechanical ventilation should be provided.

32.6.4.   The air supply should be free from contamination.

32.6.5.   Ventilation ducts should be airtight.

32.6.6.   In tunnels where blasting is done:

(a) an adequate supply of air should be taken to the face by mechanical ventilation;

(b) after every blast the face should be cleared of dust as far as practicable by exhaust ventilation; and

(c) if necessary to remove the fumes, auxiliary ventilation should be provided.

32.6.7.   Sufficient additional ventilation to prevent danger should be provided if diesel engines are used.

32.6.8.   In tunnels where there is danger from dust, the ventilation should comply with the relevant requirements of section 32.14.

32.7. Fire protection

32.7.1.   No combustible structure other than the headframe should be erected over a shaft or a tunnel mouth.

32.7.2.   No combustible structure, such as a storage building, should be built within 30 m (100 ft) of a shaft, tunnel mouth, hoisting-engine house or ventilation-fan house.

32.7.3.   No flammable oils or other highly flammable material should be stored within 30 m (100 ft) of a shaft, tunnel mouth or explosives magazine.

32.7.4. Oil escaping from tanks and drums should not be able to flow to within 30 m (100 ft) of a shaft or tunnel mouth.

32.7.5. As far as practicable, combustible materials should be kept out of tunnels.

32.7.6. No flammable liquids should be stored underground in bulk.

32.7.7. Lubricating oils, grease and rope dressings underground should:

(a) be kept in closed metal containers; and
(b) be stored in a safe place away from shafts, hoists, explosives magazines and timber.

32.7.8. Large supplies of grease or lubricating oil should not be kept underground.

32.7.9. Oily waste and rags used with machinery should:

(a) be kept in closed metal containers; and
(b) be removed to the surface at frequent intervals.

32.7.10. Waste and decayed timber should be promptly removed from the underground workings.

32.7.11. No combustible rubbish of any kind should be allowed to accumulate underground.

32.7.12. Unless there is complete freedom from fire or explosion risk, no naked lights and no smoking should be allowed underground.

32.7.13. Petrol engines should not be used underground except under conditions approved by the competent authority.

32.7.14. If welding or flame cutting is done underground:

(a) timber supports and other combustible structures or materials should be protected by a fireproof screen;
(b) suitable fire extinguishers should be kept readily available; and
(c) a constant watch should be kept for outbreaks of fire.

32.7.15. Sufficient fire-fighting equipment of suitable types should be provided in tunnelling operations.

32.7.16.   All shaft landings should be provided with:

(a) hose connections and an adequate length of hose; and

(b) suitable portable fire extinguishers.

32.8. Electricity

32.8.1.   Electrical installations in shafts and tunnels should comply with the requirements of Chapter 17 and with the requirements of this section.

Cutting off power on surface

32.8.2.   Main switchgear for cutting off the supply of electricity from all underground installations should:

(a) be installed on the surface;

(b) be accessible only to authorised persons; and

(c) be attended by a competent person authorised to operate it.

Cutting off power underground

32.8.3.   Efficient means should be provided to cut off the supply of electricity at the origin of every underground circuit.

Earthing

32.8.4.   Where the voltage exceeds safety extra-low voltage [1] (or 65 V for telephones), the following should be effectively earthed:

(a) armouring and metallic coverings of cables;

(b) external metallic parts of electrical appliances such as generators, transformers and motors which are not normally live; and

(c) metallic parts in the immediate vicinity of live conductors.

32.8.5.   Earthing systems should be so installed that no dangerous voltage can arise between earthed parts and the earth.

---

[1] As defined in section 17.1.

Lightning protection

32.8.6. Where necessary, suitable lightning arresters should be installed on the surface to protect the installation below ground from abnormal voltage due to atmospheric electricity.

Conductors

32.8.7. The main cables supplying current to electric motors installed in the vicinity of shafts (such as those for underground fans or drainage pumps) should be duplicated if the stopping of these motors would cause danger.

32.8.8. The external coverings of cables should be such that they cannot contribute to the spread of fire.

32.8.9. Cables should be heavily insulated.

32.8.10. The metal sheathing of any armoured cable should:

(a) be electrically continuous throughout its length;

(b) be earthed;

(c) be effectively protected against corrosion wherever necessary; and

(d) not be used as a current-carrying conductor.

32.8.11. Cables should be so placed and secured as to ensure maximum protection against mechanical damage of all kinds, and particularly that due to their own weight, to bending or twisting, to traffic or to ground movements.

32.8.12. Supporting devices for cables should:

(a) be of adequate mechanical strength;

(b) permit the cable to move when it is subjected to any unusual pull, except in shafts; and

(c) be spaced sufficiently close to prevent dangerous sagging.

32.8.13. Electric conductors should not be supported on spikes or other makeshift supports.

32.8.14. Cables in shafts should have an armouring strong enough to support, without dangerous sagging, their weight, taking into account the spacing of supports.

32.8.15. The armouring referred to in paragraph 32.8.14 should have a stress safety factor of at least 3.

Switches

32.8.16. Switches should be of the enclosed safety type.

Fixed lighting

32.8.17. Fixed lamps underground should be fitted with a strong protective cover of glass or other transparent material.

32.8.18. If the cover is not highly resistant to impact, it should be provided with a guard.

32.8.19. Whenever required by local conditions, the lamp fitting should be proof against dust and water.

32.8.20. Fixed lighting apparatus should be so constructed that the lamp bulb or tube can be replaced without danger of electric shock, unless there are sufficient switches for all supply conductors in the circuit to be made dead.

32.8.21. When the installation is supplied from a trolley wire system, the lighting should be limited to the parts of the roadway containing the trolley wire and the immediate vicinity of those parts.

32.8.22. When the installation is supplied from a trolley wire system, each lamp should:

(a) be protected by a fuse inserted in the circuit between the lamp and the trolley wire;

(b) have an earthing conductor separate from the return conductor; and

(c) be sufficiently well insulated from any metal road support.

Handlamps

32.8.23. The voltage of handlamps (portable lamps) used underground should not exceed safety extra-low voltage.

Transport of workers

32.8.24. The conveyance of persons in trains hauled by electric trolley locomotives should only be permitted under conditions to be specified by the competent authority.

32.8.25. Unless other effective safeguards are provided, the conveyance of persons should be permitted only in cars with well-earthed roofs adequately protecting passengers against contact with live conductors.

32.8.26. At all stations where persons enter or leave roofless cars:

(a) there should be a switch by means of which the supply of electricity can be switched off from the trolley wire throughout the length of the station;

(b) there should be light signals to indicate whether the trolley wire is live or dead, so arranged that at least one can be seen from any part of the train; and

(c) adequate fixed lighting should be provided.

32.9. Underground lighting

32.9.1. All places where workers have to work or pass should be adequately lit.

32.9.2. In addition to the main lighting, there should be emergency lighting that functions long enough to enable the workers to reach the surface safely.

32.9.3. All hoisting, pumping and other machinery should be sufficiently lit to enable moving parts to be readily distinguished.

32.9.4. Floodlights in tunnels should:

(a) only be installed at places at least 3 m (10 ft) high; and

(b) shine through frosted glass.

32.9.5. Workers should not enter unlit workings without a portable light.

32.10. Drilling

32.10.1.   When drilling is done in rock, loose rock should be scaled down to protect drillers against falls of ground; where this is not practicable, a protective canopy or overhead screen should be provided.

32.10.2.   High drilling rigs should be provided with:

(a) safe means of access such as a ladder or stairs complying with the relevant requirements of Chapter 4;

(b) guard-rails and toe-boards complying with the requirements of paragraphs 2.6.1 to 2.6.5; and

(c) suitable storage accommodation for drill steels, such as racks or boxes.

32.10.3.   Air hoses should be secured by chains, or have self-locking couplings, or otherwise be prevented from causing danger if a coupling fails.

32.10.4.   Drillers should wear goggles and heavy gloves.

32.10.5.   Precautions against dust in drilling should be taken in compliance with the requirements of paragraphs 32.14.9 to 32.14.12.

32.11. Transport, storage and handling of explosives

32.11.1.   The transport, storage and handling of explosives should comply with the requirements of Chapter 22 and with the requirements of this section.

32.11.2.   Explosives should not be conveyed in a shaft cage or bucket together with other materials.

32.11.3.   Explosives and detonators should not be conveyed together in a shaft unless they are in a suitable powder car.

32.11.4.   If explosives are conveyed to the face by rail:

(a) explosives should not be in the same car as detonators; or

(b) a suitable powder car should be used; and

(c) not more than the amount of explosive required for one shift should be placed in the powder car.

32.11.5. Powder cars should have separate compartments for explosives and detonators.

32.11.6. Both compartments should be adequately insulated from the car frame and other conducting material.

32.12. Blasting

32.12.1. Only electric blasting should be allowed in tunnels.

32.12.2. Electric blasting operations in tunnels should comply with the relevant provisions of Chapter 23.

32.12.3. Blasting circuits should be independent of any power, lighting or other circuit.

32.12.4. Blasting circuits should be tested before detonators are attached.

32.12.5. No other electrical circuit should be installed on the same side of the tunnel as the blasting circuit.

32.12.6. Before any shot is fired, all electrical circuits other than the blasting circuit should be de-energised for an adequate distance from the firing point.

32.12.7. Only suitable battery lamps should be used for loading shotholes.

32.12.8. After every blast, the sides and roof should be inspected and cleared of loose rock.

32.12.9. Precautions against dust in blasting should be taken in compliance with the requirements of paragraphs 32.14.13 to 32.14.15.

32.12.10. When an electrical storm is approaching, blasting operations and preparations for them should be discontinued.

32.12.11. Arrangements should be made to warn blasting crews of the approach of electrical storms.

32.13. Haulage

General provisions

32.13.1.   Rail tracks and rolling stock of mechanical haulage systems should comply with the relevant requirements of section 9.1.

32.13.2.   Electric trolley haulage should comply with the requirements of section 17.6.

32.13.3.   In tunnels where there are rail tracks, unless there is adequate clearance between the rolling stock and the sides, recesses should be provided at suitable intervals.

32.13.4.   Recesses referred to in paragraph 32.13.3 should be large enough to accommodate two persons and should be at least 60 cm (2 ft) deep.

32.13.5.   No locomotive producing smoke should be allowed in tunnels.

32.13.6.   Cars for the transportation of workers should be provided with seats.

32.13.7.   When cars are moved by hand, they should be provided with handles protecting the workers' hands.

Operations

32.13.8.   Mechanical haulage operations should be governed by signals in conformity with the relevant requirements of section 1.8.

32.13.9.   A warning signal should be given:

(a) before a train leaves or arrives at a station;
(b) when a train approaches a curve; and
(c) when otherwise necessary to prevent danger.

32.13.10.   Trains and single cars entering unlit workings should have headlights and tail lights.

32.13.11.   Diesel engines for driving locomotives should be shut off whenever the locomotive stops.

32.13.12.   Standing cars should be blocked to prevent them from running away.

32.13.13.   Derailed cars should be rerailed with levers, jacks or cranes.

32.13.14.   Rerailing by hauling with a winch should only be done under the control and supervision of a competent person.

32.13.15.   On inclines, derailed cars should be blocked while being rerailed.

32.13.16.   Cars moved by hand should be pushed, not pulled.

32.13.17.   Workers should not be transported on locomotives or in cars other than those specially provided for man trips.

32.13.18.   Explosives should be transported in conformity with the requirements of section 32.11.

32.14. Dust

General provisions

32.14.1.   Adequate measures should be taken to prevent the formation of, or to suppress, all dust in tunnelling operations.

32.14.2.   Particular care should be taken to prevent the formation of, or suppress, siliceous dusts consisting of particles less than 5 microns in size.

32.14.3.   Dust should be suppressed as close as possible to its source.

32.14.4.   The ventilation in tunnelling operations should:

(a) supply air that is as clean as practicable to the workplaces;
(b) effectively dilute and remove airborne dust; and
(c) not have a velocity high enough to raise dust.

32.14.5.   The air in tunnelling operations should be sampled for dust at suitable intervals by competent persons.

Water

32.14.6.   If water is used for the prevention and suppression of dust, an adequate supply of water should be available in tunnelling operations.

32.14.7. Water used for dust-suppression purposes should not present any health risk.

32.14.8. Water used for consolidating dust should not be applied with such force as to raise dust into the air.

Drilling

32.14.9. If drilling in rock is done dry, the dust produced should be effectively exhausted and collected.

32.14.10. If drilling in rock is done wet, the drill should be so constructed that it cannot be operated unless the water feed is operating.

32.14.11. In wet drilling, the water should penetrate to the bottom of the hole in sufficient quantity and under sufficient pressure to render the dust harmless.

32.14.12. Pneumatic drills with an axial water feed should not carry air with the water through the drill steel.

Blasting

32.14.13. The times for blasting should be chosen so that the smallest practicable number of workers are exposed to the dust produced.

32.14.14. Before any shots are fired, the floor, roof and sides in the vicinity should be thoroughly wetted, if practicable.

32.14.15. The dust from blasting should be actively removed by ventilation and, where necessary, be allayed with sprays or fog guns, or passed through filters.

Transport

32.14.16. Loose rock should be adequately wetted during loading, transport and unloading underground.

32.14.17. Excavated material should not be exposed to high-velocity air currents during transport.

32.14.18. Either transfer and loading points should be so designed as to prevent the dispersal of dust into the air, or the dust produced should be suppressed by suitable wet or dry methods.

32.14.19. Spillage occurring during transport should be constantly cleaned up.

32.14.20. Adequate precautions should be taken to reduce the formation and spillage of dust on conveyors as much as practicable.

32.14.21. Fine dust adhering to the belts of belt conveyors should be removed continuously and collected.

32.14.22. The movement of conveyors should, as far as practicable, be so regulated that no material will accumulate at transfer points.

Stone crushing

32.14.23. If any stone-crushing equipment is used underground, adequate measures should be taken to prevent any dust from it penetrating to areas occupied by workers.

Airborne dust

32.14.24. Airborne dust should be precipitated, filtered or removed to the outside air.

32.14.25. Extracted dust should be removed in dust-tight containers or in the form of mud after wetting.

32.14.26. Filters should be cleaned at suitable intervals to maintain them in an effective working condition.

32.15. Underground pipelines

32.15.1. Adequate ventilation should be provided for workers in pipelines.

32.15.2. When laying pipes in water-bearing ground, a flood gate should be provided at the end section.

32.15.3. When high-pressure pumps are used:

(a) they should be tested before they begin to work;

(b) the pressure gauge should be kept under observation; and

(c) the maximum safe working pressure should not be exceeded.

32.15.4. When bodies of water or explosive gases may be encountered, trial boreholes should be drilled ahead of the workings.

32.15.5. There should be reliable means of communication between workers inside pipes and persons outside.

32.15.6. It should be possible for workers employed in piping to reach a safe place quickly in an emergency.

32.15.7. Workers installing pipelines should not work beyond the excavating blade of boring equipment.

32.15.8. Where boring equipment is used, pipe should be laid as close as possible to the excavating blade so that workers will not be exposed to the danger of collapsing earth.

32.15.9. Adequate arrangements should be made to rescue workers who are in danger and cannot reach a safe place.

33. Work in compressed air caissons and tunnels

33.1. General provisions

33.1.1. No person should be subjected to a pressure exceeding 3.5 kg/cm² (50 lb/in²) except in emergencies.

33.1.2. For every shift a record should be kept showing the time every worker spends in the working chamber and the time taken for decompression.

Personnel

33.1.3. No person should be employed in compressed air unless:

(a) he has experience of such work; or

(b) he is under the constant supervision of an experienced person.

33.1.4. No person under 20 years of age should be employed in compressed air.

33.1.5. When beginning employment in compressed air every worker should be given a leaflet setting out the precautions that he should observe before, during and after work.

Medical supervision

33.1.6. No person should be employed in compressed air unless he has been medically examined and found fit for such employment.

33.1.7. If the air pressure exceeds 1.25 kg/cm² (18 lb/in²) the medical examination should have been made within the four weeks preceding his employment.

33.1.8. Workers who have been employed continuously in compressed air at a pressure less than 1.5 kg/cm² (21 lb/in²) should be medically re-examined every two months; if the pressure is higher, the period between re-examinations should be shorter.

33.1.9. Workers who have been absent from work in compressed air for any period due to illness or for ten days or more for reasons other than illness should be medically re-examined.

33.1.10. Persons employed in compressed air should be under competent medical supervision.

33.1.11. For every project on which workers are employed in compressed air, a physician and a registered nurse or a qualified first-aid attendant familiar with compressed air work should be appointed.

33.1.12. The physician should be readily available at all times.

33.1.13. The nurse or the qualified first-aid attendant should be in attendance in the first-aid room while work in compressed air is going on.

33.1.14. Superintendents, foremen, and a sufficient number of workers, at least one in each team, should be competent to administer first aid.

33.1.15. A first-aid room complying with the requirements of section 38.2 should be provided near the entrance shaft.

33.1.16. In caissons there should be a first-aid box in the working chamber.

33.1.17. In tunnels with a bulkhead there should be a first-aid box on each side of the bulkhead near the man-lock entrance.

33.1.18. When persons are employed in compressed air at a pressure exceeding 1.25 kg/cm² (18 lb/in²) the employer should inform a neighbouring hospital of the position of the work site and of the name and address of the physician exercising medical supervision.

33.1.19. Every person employed in compressed air at a pressure exceeding 1.25 kg/cm² (18 lb/in²) should be provided with a badge or label to be worn next to the body indicating that he has been employed in compressed air and giving the address of the medical lock at his place of employment.

33.1.20. The identification badge should state that the wearer should be taken to the medical lock and not to a hospital if he is ill.

Hygiene, welfare

33.1.21. For the use of persons employed in compressed air there should be provided:

(a) accommodation for clothing and facilities for changing complying with the requirements of section 37.7;

(b) facilities for washing complying with the relevant requirements of section 37.5; and

(c) adequate and suitable facilities for remaining on the site after decompression, including shelters with seats.

33.1.22. Toilet facilities complying with the relevant provisions of section 37.4 should be provided at a suitable place outside the air chambers.

33.1.23. Where practicable at least one chemical toilet should be provided in the working chamber.

33.1.24. No person should commit a nuisance underground or on the surface.

33.1.25. When persons are employed in compressed air at a pressure exceeding 1.25 kg/cm^2 (18 lb/in^2) they should be supplied with hot drinks when they leave the man lock and while they are in the medical lock.

33.1.26. No person should consume alcohol while in compressed air, or before going into compressed air.

33.1.27. No person should smoke in compressed air.

33.1.28. All parts of caissons and facilities for the workers employed in them should be kept in a clean and sanitary condition.

Compression, decompression

33.1.29. Compression and decompression should be carried out in accordance with requirements which should be laid down by the competent authority.

33.1.30. During compression the pressure should not be raised to more than about 0.35 kg/cm² (5 lb/in²) until the lock attendant has ascertained that no person is complaining of discomfort, and thereafter it should be raised at a rate not exceeding about 0.7 kg/cm² (10 lb/in²) per minute.

33.1.31. If during compression any person is suffering from discomfort, compression should stop and the pressure be gradually reduced.

33.1.32. Decompression should be carried out in conformity with timetables laid down in national or other official regulations respectively for normal and for phase decompression.

Hours of work

33.1.33. The length of shifts and rest periods should be specified in national or other official regulations in accordance with the air pressure.

33.2. Work in caissons

Means of access and egress

33.2.1. Safe means of access should be provided to every place where persons are employed in a cofferdam or caisson.

33.2.2. Adequate means for persons to reach places of safety should be provided in every cofferdam and caisson in case of an inrush of water.

33.2.3. Where practicable, a stairway complying with the requirements of section 4.7 should be provided in shafts.

33.2.4. Where stairways are impracticable, ladders complying with the relevant requirements of sections 4.1 and 4.6 should be provided.

Shaft and caisson construction

33.2.5. When necessary to prevent danger, caissons and shafts should:

(a) be adequately braced; and
(b) be firmly secured in position.

33.2.6. The bracings and ties of caissons should be adequately secured to prevent their accidental displacement.

33.2.7. Locks and shafts should be constructed of steel or other suitable metal of adequate thickness.

33.2.8. Before being taken into use shafts should undergo an adequate hydrostatic or air-pressure test at 5.25 kg/cm^2 (75 lb/in^2).

33.2.9. Every shaft and caisson containing flammable material should be provided with a water line, sufficient hose connections and sufficient hose or appropriate extinguishers.

Working chambers

33.2.10. Every working chamber should be provided with a wet-bulb thermometer.

33.2.11. Work under pressure when the wet-bulb temperature exceeds 28 °C (80 °F), should be restricted unless it is absolutely necessary.

33.2.12. While any person is in a working chamber, the door between the chamber and a man lock leading to a lower pressure should as far as practicable be kept open if the lock is not in use.

Medical locks

33.2.13. Where the pressure in a working chamber ordinarily exceeds 1.25 kg/cm^2 (18 lb/in^2), a suitable medical lock should be provided for the treatment of workers employed in compressed air.

33.2.14. The medical lock should have two compartments so that it can be entered under pressure.

33.2.15. Medical locks should be adequately ventilated, heated and lit.

33.2.16. Medical locks should be provided with suitable equipment including a couch, blankets, dry woollen garments, a food locker, means of communication and signalling to the outside and between the compartments, and windows by which persons in either compartment can be observed from the outside.

33.2.17. Medical locks should be kept ready for use at all times.

33.2.18. While any person is employed in compressed air a medical lock should be in the charge of a suitably qualified person.

Man locks

33.2.19. Every man lock should be of adequate internal dimensions for the purpose for which it is used.

33.2.20. Every man lock should be equipped with:

(a) pressure gauges that:

    (i) indicate to the man lock attendant the pressure in the lock and in each working chamber to which it affords direct or indirect access; and

    (ii) indicate to the persons in the lock the pressure in it;

(b) a clock or clocks so placed that the lock attendant and the persons in the lock can readily ascertain the time;

(c) efficient means of verbal communication between the lock attendant, the lock and the working chamber or chambers;

(d) means of enabling the persons in the lock to convey visible or other non-verbal signals to the lock attendant; and

(e) efficient means enabling the lock attendant, from outside the lock, to reduce or cut off the supply of compressed air to the lock.

33.2.21. In every man lock there should be a suitable notice indicating the precautions to be taken by persons during compression and decompression, and after decompression.

33.2.22. Man locks should only be used for compression and decompression of persons and not for the passage of plant and material other than hand tools.

33.2.23. Man locks should be kept clean and suitably warm.

33.2.24. Every man lock should, while any person is in it or in any working chamber to which it affords direct or indirect access, be in the charge of an attendant who should:

(a) control compression and decompression in the lock; and

(b) if the pressure exceeds 1.25 kg/cm² (18 lb/in²), keep a register showing:

   (i) the times at which each person enters and leaves the lock;

   (ii) the pressures at the times of entering and leaving; and

   (iii) the times taken to decompress each person.

Air supply

33.2.25. Compressed-air installations should be provided with air-supply plant capable of supplying any working chamber with sufficient fresh air at the pressure in the chamber, and not less than 0.3 m³ (10 ft³) per minute per person in the chamber.

33.2.26. Pollution of the air fed in the caisson from a compressor or any other source should be carefully prevented.

33.2.27. All air lines should be in duplicate and be equipped with non-return valves that will prevent air from escaping from the working chamber into the air line if the pressure in it falls.

33.2.28. There should be a sufficient reserve of air in compressor installations to allow a safe margin for breakdowns or repairs.

33.2.29. There should be a stand-by or reserve compressor for emergencies.

33.2.30. Two separate power units supplied from independent sources should be provided for each compressor.

33.2.31. Valves or taps for controlling the air flow should enable it to be controlled accurately.

33.2.32. Persons in the lock should not be able to reduce the air pressure except:

(a) under the control of the lock attendant; or

(b) in an emergency, by special means that should normally be kept sealed or locked.

33.2.33. Working chambers should be provided with exhaust valves for clearing the air when necessary, as after a blasting operation.

Signalling

33.2.34. Reliable means of communication such as bells, whistles or telephones should be maintained at all times between the working chamber and all surface installations.

33.2.35. Signalling arrangements should comply with the requirements of section 1.8.

33.2.36. The code of signals should be conspicuously displayed in convenient positions at workplaces.

33.2.37. Before any cage, skip, bucket or elevator is moved the starting signal should be repeated by the persons to whom it is sent.

Lighting

33.2.38. All locks and working chambers should be provided with adequate electric lighting.

33.2.39. There should be two separate lighting installations supplied from independent sources of current.

33.3. Work in tunnels in compressed air

33.3.1. Work in tunnels in compressed air should comply with the requirements of section 33.2 and with the requirements of this section.

Bulkheads

33.3.2. The bulkhead separating the working chamber from areas of lower pressure should be of sufficient strength to withstand safely the maximum pressure to which it will be subjected.

33.3.3. When necessary to prevent danger in the event of rapid flooding, the bulkhead should be sufficiently close to the face or shield to allow the workers to escape in an emergency.

Safety curtains

33.3.4. Safety curtains should be provided within 60 m (200 ft) of the working face in all tunnels where there is a danger of an inrush of water or material.

33.3.5. Safety curtains should:

(a) be made of incombustible material;
(b) be installed in the crown of the tunnel;
(c) provide an airtight seal with the lining of the tunnel;
(d) be adequately braced and reinforced; and
(e) extend to the centre line of the tunnel or other safe distance.

Air supply

33.3.6. If the compressor is driven by electricity, stand-by compressor plant should be provided capable of maintaining at least 50 per cent of the air supply if the electrical power fails.

33.3.7. If the compressors are not driven by electricity, not more than half of them should be driven from any one power source.

33.3.8. Each air line should be equipped with an adequate air receiver, a stop valve, a pressure-reducing valve, and a non-return valve close to the man locks.

33.3.9. The air supply should be provided by duplicate air lines between the air receiver and the working chamber.

33.3.10. An adjustable safety valve should be fitted on the outside of the bulkhead to a separate pipe leading from the working chamber through the bulkhead to the outside air.

33.3.11. An oil separator should be provided between the air compressor and the air receiver, and a filter between the air receiver and the working chamber.

33.3.12. After compression the air should be cooled if necessary to keep the temperature of the working chamber below 28 °C (80 °F) wet bulb.

Man locks

33.3.13.   Where practicable, in addition to a man lock and a material lock, tunnels should have an emergency lock capable of holding an entire heading shift.

33.3.14.   Emergency man locks should be kept open towards the face, and be ready for use at all times.

33.3.15.   Emergency man locks should be installed in the crown of the tunnel.

33.3.16.   Man locks should be installed as high in the tunnel as is practicable.

Medical locks

33.3.17.   A medical lock complying with the relevant provisions of section 33.2 should be provided when work in compressed air is carried on in tunnels at pressures exceeding 1.25 kg/cm^2 (18 lb/in^2).

Gangways

33.3.18.   In all tunnels 5 m (16 ft) or over in diameter or height an overhead gangway should be provided from the working surface to the nearest airlock with an overhead clearance of at least 1.80 m (6 ft).

33.3.19.   Overhead gangways should be provided with guard-rails and toe-boards complying with the requirements of paragraphs 2.6.1 to 2.6.5.

Fire protection

33.3.20.   Every tunnel should be provided with a water line extending into the working chamber to within 30 m (100 ft) of the working face, sufficient hose connections at suitable places, and sufficient hose.

33.3.21.   If there is a danger of fire a suitable hose should be provided on both sides of the tunnel bulkhead.

33.3.22.   In areas under pressure, walkways, stairways and ladders should be of incombustible material.

Rail haulage

33.3.23.   An automatic stop block or derailing device should
be provided:

(a) at the top of every haulage incline; and

(b) at a suitable distance upgrade from any point where runaway
   cars could damage the shield or the airlock.

33.3.24.   Holding devices should be used while cars are being
loaded.

Blasting

33.3.25.   When blasting work is being done in compressed
air in tunnels:

(a) no worker other than the blaster and his assistants should be
   in a working chamber while boreholes are being loaded; and

(b) no worker should re-enter a working chamber after a blast
   until the fumes have cleared.

34. Diving

34.1. General provisions

34.1.1.   The following provisions are intended to apply to diving with specialised suits or helmets.   Diving with self-contained, underwater breathing apparatus is of a specialised nature and is not covered in this Code.

Personnel

34.1.2.   No person under the age of 20 or over the age of 55 should be employed as a diver.

34.1.3.   No person should be employed as a diver unless he has experience of the work or is being trained under the supervision of an experienced diver.

34.1.4.   No diving operations should be undertaken unless a team consisting of two divers, one attendant (signalman) and one or more pumpmen is provided.

34.1.5.   In all diving operations a second diver with a complete set of diving equipment should be constantly available for employment in an emergency.

34.1.6.   On boats used for diving operations there should be at least one properly qualified sailor.

Medical supervision

34.1.7.   No person should be employed as a diver unless he has been medically examined within the previous six months and been found fit for such employment.

34.1.8.   If by reason of disease or injury a diver is incapacitated for employment as a diver for more than two weeks he should not be employed again as a diver until he has been medically examined and found to be fit for such employment.

34.1.9.   Divers should be medically re-examined at appropriate intervals as determined by the competent authority.

34.1.10. Unless authorised by a doctor, divers should not be employed at a depth exceeding 10 m (33 ft) or in dangerous operations.

34.1.11. Divers should immediately report any indisposition to a doctor, first-aid attendant or supervisor.

34.1.12. If a diver meets with an accident under water:

(a) he should receive medical attention as soon as practicable;

(b) if he has ascended too quickly he should be put under the appropriate pressure under medical supervision, in an air lock.

34.1.13. At a suitable position at the workplace, notices should be displayed giving:

(a) the name, address and telephone number of the nearest doctor familiar with diving conditions;

(b) the name, address and telephone number of the nearest available diver; and

(c) the address and telephone number of the nearest recompression chamber.

34.1.14. If extensive diving operations are carried on in deep water a medical lock complying with the relevant requirements of section 33.2 should be provided.

Hours of work

34.1.15. The hours of work and rest periods of divers should:

(a) be suited to their physical fitness and underwater depths and pressures; and

(b) conform to scales that should be laid down in national or other official regulations.

34.2. Provision of diving equipment

Dress

34.2.1. Divers should be provided with adequate diving equipment including means of access to and from the water, and means of communication and a lifeline with an adequate belt.

34.2.2.   Separate sets of warm clothing, woollens or flannels should be provided for each diver.

34.2.3.   If necessary to prevent danger in cold water the diver should be equipped with woollen undersuit, helmet and gloves.

Means of entering and leaving the water

34.2.4.   For entering and leaving the water the diver should be provided with equipment such as steps, a ladder with a hand rope, or a platform.

Air supply

34.2.5.   If a diver is supplied with air through a pipeline the diving equipment should include air pumps and compressors or cylinders.

34.2.6.   If air compressors are used a sufficient reserve of air should be provided to enable the diver to reach the surface if the compressors fail.

34.2.7.   While under water divers should be provided with respirable air in adequate quantity and at a pressure suitable for the diving equipment and the working conditions.

34.2.8.   As long as the diver is wearing his diving suit or diving helmet the pumps should be kept in operation.

34.2.9.   The air line between the compressor and the diver should:

(a) be made of rubber reinforced with webbing or of equivalent construction;

(b) be capable of withstanding without deformation the highest hydraulic pressure to which it will be subjected; and

(c) possess adequate breaking strength.

34.2.10.   The resistance to hydraulic pressure and the breaking strength of the air line referred to in paragraph 34.2.9 should have been verified by pressure and tensile tests.

34.2.11.   Joins in air lines should be made by screw couplings secured against coming apart.

34.2.12. The air line should be equipped with:

(a) an air receiver;

(b) oil and water filters;

(c) a safety valve;

(d) a stop valve;

(e) a reducing valve; and

(f) a pressure gauge.

34.2.13. The air receiver should be large enough to ensure an adequate air supply to the diver if the compressor fails until the regular supply can be restored by a hand pump, reserve compressor or other effective means.

34.2.14. There should be a non-return valve between the air receiver and compressor.

34.2.15. If the air pump is power-driven:

(a) it should be possible to convert it quickly for manual operation; or

(b) a manual pump of suitable capacity and producing a sufficient pressure should be kept readily available.

34.2.16. Where divers are using pneumatic tools, air for the tools should be taken from a source entirely separate from the divers' air supply, for example, from a separate air receiver.

Lifelines

34.2.17. The lifeline should:

(a) be of good quality rope;

(b) have an adequate breaking strength as verified by a tensile test; and

(c) be of adequate length for the operations in which it is used.

Signalling and communication

34.2.18. Diving operations should be governed by signals in accordance with accepted standards.

34.2.19. The attendant (signalman) should be responsible for the safety and the proper working of the signal and air lines.

34.2.20. When diving operations are conducted from land, signals should be flown at adequate distances above and below the point of operation.

34.2.21. When diving operations are conducted in a fairway, appropriate signals should be flown.

34.2.22. When diving operations are conducted from a boat, the boat should fly signals to indicate this.

34.2.23. In dangerous operations such as blasting underwater and working in a swift current, there should be telephonic communication between the diver and the surface.

34.2.24. The diver's telephone should be so designed that he does not have to hold it in his hands.

Lighting

34.2.25. For work at night or in dark places underwater, electric lighting should be provided.

34.2.26. Hand lamps for divers should be:

(a) battery lamps; or
(b) lamps supplied at a voltage not exceeding 24 V a.c.; or
(c) lamps with reinforced insulation.

Boats, rafts, etc.

34.2.27. When diving operations are conducted from a moored vessel or from land, a suitable boat provided with rescue equipment such as gaffs and ropes should be kept in readiness.

34.2.28. Boats, rafts, platforms, etc. from which diving operations are carried on should:

(a) have adequate space for working and the storage of material and equipment;
(b) possess adequate stability; and
(c) be securely anchored having regard to the prevailing wind and current conditions.

First-aid and rescue equipment

34.2.29. At places where diving operations are carried on, first-aid and rescue equipment complying with the relevant provisions of section 38.2 should be provided.

34.3. Inspection, testing and care of diving equipment

34.3.1. No diving equipment should be taken into use for the first time unless it has been thoroughly inspected and tested and found to be safe to use.

34.3.2. Diving equipment should be thoroughly inspected at least once in three months.

34.3.3. No air pump, compressor, cylinder or pipeline should be used in any diving operation unless within the previous 24 hours it has been tested for leakage and a pressure exceeding that at which the diver has to descend is maintained for a sufficient period when the pump or compressor is not working.

34.3.4. No diving equipment should be used unless any inlet and outlet valves on the diver's dress, and any demand or regulator valve on the equipment, have been inspected within the previous 24 hours and found to be in safe working order.

34.3.5. No self-contained diving equipment should be used unless it has been tested within the previous 24 hours and found to be functioning efficiently.

34.3.6. All diving dresses and equipment should be kept when not in use in a suitable room or locker which should not be used for any other purpose.

34.3.7. All diving dresses and equipment should be cleaned with fresh water, drained and dried before being stored in the room or locker provided.

34.3.8. Diving dresses and air pipes should be kept free from condensed moisture and contact with grease, oil or tar.

34.4. Diving operations

General provisions

34.4.1. Before diving operations begin, the diver should familiarise himself with current conditions, traffic, and dangers such as underwater cables, suction pipes, and hawsers.

34.4.2. Divers should not enter the water if:

(a) they are suffering from any illness or are feeling unwell;
(b) they are under the influence of alcohol;
(c) they are fatigued;
(d) they have just had a heavy meal; or
(e) they have no food in the stomach.

34.4.3. When the diver enters the water:

(a) he should use the ladder or other equipment provided for the purpose and not jump in;
(b) the attendant (signalman) should make sure that the diving suit is watertight;
(c) the attendant (signalman) should lower him slowly;
(d) if he sinks too fast the attendant (signalman) should stop him, and if he fails to signal, bring him to the surface.

34.4.4. So long as the diver is in the water he should be under the constant supervision of an attendant (signalman).

34.4.5. In particular the attendant (signalman) should:

(a) watch the air and lifelines;
(b) exchange signals with the diver at suitable intervals;
(c) ensure that the diver is not endangered by activities in the vicinity; and
(d) bring the diver to the surface if he is endangered.

34.4.6. In freezing weather, precautions should be taken to prevent danger to the diver from ice in the air line, valves, etc.

34.4.7. When necessary to prevent danger, precautions should be taken to prevent vessels from approaching a diver in the water.

34.4.8.   When diving is done from a boat or other floating equipment, the equipment should be securely anchored before diving operations begin.

34.4.9.   When a diver is lowered by a derrick or other lifting appliance, the appliance should not be used for any other purpose while the diver is attached to it.

34.4.10.   The operation of the appliance referred to in paragraph 34.4.9 should be governed by the diver's attendant (signalman).

34.4.11.   While a diver is working from a ship, adequate precautions should be taken to prevent any movement by the screws or the rudder, and any discharge from underwater valves.

34.4.12.   In salvage operations no material should be raised until the diver has given the appropriate signals:

(a) for taking the strain; and

(b) after he has moved to a safe place, for lifting to the surface.

34.4.13.   If the signal and air lines cannot be kept clear when heavy or bulky loads such as sheet piling, pipes and metal sheets are being raised or lowered the diver should be brought to the surface.

34.4.14.   So long as the diver is in the water nothing should be thrown down or conveyed over the workplace.

34.4.15.   If the diver rises too quickly:

(a) the signal and air lines should be rapidly hauled in;

(b) the diver should be lowered again as soon as practicable and then raised after a short interval.

34.4.16.   Divers should be brought to the surface slowly and in stages conforming to scales that should be laid down in national or other official regulations.

Underwater blasting

34.4.17.   Underwater blasting should only be done under the supervision of a qualified blaster.

34.4.18. Explosives and ignition equipment used in underwater blasting should be so constituted as to remain waterproof for the duration of their immersion in water.

34.4.19. Only low-tension submarine detonators should be used.

34.4.20. Electric leads and fuses should be protected against damage due to the force of waves.

34.4.21. Blasting cables should be effectively insulated two-core cables.

34.4.22. The insulation of connections in electric leads should be watertight.

34.4.23. Precautions should be taken to ensure that divers are not exposed to danger from:

(a) objects striking against blasting equipment;

(b) dragging of leads or fuses;

(c) kinks in leads or fuses; and

(d) entangling of the air line or lifeline in leads or fuses.

34.4.24. Charges should only be prepared and detonators attached on the diving vessel or on land.

34.4.25. The blasting cable should be firmly secured at the blasting position, by tying, weighting or other effective means.

34.4.26. The blasting position should be made clearly recognisable from the surface.

34.4.27. Only one blasting cable should be used and only one shot at a time should be connected to the cable.

34.4.28. No shot should be fired:

(a) before the diver is out of the water;

(b) until it has been made sure that the charge has not been accidentally displaced by the diver;

(c) until all boats, rafts, etc. are at a safe distance; and

(d) until all other necessary precautions have been taken.

34.4.29. Blasting under ice should comply with the relevant requirements of paragraphs 34.4.17 to 34.4.28.

34.4.30. Charges under ice should be prevented from being displaced by water currents.

Underwater welding and flame cutting

34.4.31. Underwater welding and flame cutting should only be done by persons who:

(a) are properly qualified divers; and

(b) are competent to do the work.

34.4.32. Special equipment for underwater welding and flame cutting should be of types approved by the competent authority.

34.4.33. When work is done on containers, hollow objects and the walls of confined spaces, precautions should be taken to prevent the formation of explosive mixtures inside.

34.4.34. When work is done with oxygen and liquid fuel such as petrol, precautions should be taken to prevent fires on the surface being caused by fuel escaping from the burner.

34.4.35. Containers for fuel should be constructed and handled in conformity with the relevant requirements of section 21.2.

34.4.36. When electric-arc welding or cutting is done:

(a) the welder's dress and helmet should be adequately insulated inside; and

(b) the welder should wear insulating gloves.

34.4.37. Only direct current should be used for underwater welding and cutting.

34.4.38. It should be possible for the welder to have the current cut off immediately from the welding generator by the attendant (signalman) for operational reasons or in an emergency.

34.4.39. All conductors, connections, electrodes and electrode holders should be adequately insulated with waterproof material.

34.4.40. Electrodes should only be changed when the electrode holder has been made dead and the attendant (signalman) has so informed the welder.

35. Handling materials

35.1. Manual lifting and carrying

35.1.1. Where reasonable and practicable, mechanical appliances should be provided and used for lifting and carrying loads.

35.1.2. Workers required to handle loads should be instructed how to lift and carry safely.

35.1.3. In particular, the strain of lifting should be taken by the legs and not by the back.

35.1.4. No person should be employed to lift, carry or move any load which, by reason of its weight, is likely to jeopardise his health or safety.

35.1.5. No adult male worker should lift or carry a load exceeding 50 kg (110 lb) in weight.

35.1.6. No male young person should lift or carry a load exceeding 20 kg (44 lb) in weight.

35.1.7. No female young person should lift or carry a reasonably compact load exceeding 15 kg (33 lb) in weight.

35.1.8. No adult woman should lift or carry a load exceeding 20 kg (44 lb) in weight.

35.1.9. No pregnant woman, and no young person under the age of 16, should lift or carry loads.

35.1.10. When long objects such as poles are being carried by a number of workers;

(a) the last worker should give the signals for lifting and dropping;
(b) all the workers should lift and drop the object simultaneously on being given the appropriate signals;
(c) all the workers should be on the same side of the object;
(d) when crossing slopes the workers should be on the uphill side of the object; and
(e) the object should not be thrown down over the head.

35.2. Stacking and piling

General provisions

35.2.1. Materials and objects should be so stacked and unstacked that no person can be injured by materials or objects falling, rolling, overturning, falling apart or breaking.

35.2.2. Persons not directly concerned with the operations should keep out of areas where stacking and unstacking operations are in progress.

35.2.3. Workers should not climb on to stacks while stacking and unstacking operations are in progress.

35.2.4. Safe means of climbing stacks, such as ladders, should be provided for workers who have to climb them.

35.2.5. No worker should be employed out of sight of other workers in stacking or unstacking operations.

35.2.6. If workers have to pass regularly along the tops of stacks and from one stack to another, safe walkways and bridges protected by guard-rails or toe-boards complying with the requirements of paragraphs 2.6.1 to 2.6.5 should be provided.

35.2.7. Material should not be placed or stacked near the edge of any excavation, shaft, pit or other opening in the ground or floor so as to cause danger to any person.

35.2.8. Before starting to take down stacks, workers should clear the ground around them of any obstacles, snow, etc. that might cause danger.

35.2.9. Stacks should only be taken down from the top, and no object should be pulled out from under another.

35.2.10. Stacks that lean heavily, become unstable or threaten to collapse should only be taken down under competent supervision.

Lumber

35.2.11. Lumber should be stored on supports above the ground.

35.2.12.   Supports should be level and rest on firm ground.

35.2.13.   Layers of lumber should be separated by strips laid crosswise.

Bags of cement, lime or other similar materials

35.2.14.   Cement, lime or other similar materials in bags should:

(a) have the mouths of the bags placed inwards;

(b) have the four lowest corner bags cross tied;

(c) have a step back of one bag at every fifth row.

35.2.15.   Piles should not be more than ten bags high unless they are in an enclosure or are otherwise adequately supported.

35.2.16.   When bags are removed from the pile they should be so taken that the height of the pile remains uniform.

35.2.17.   Bags of lime, cement or other similar materials should be stored in a dry place.

Bricks, tiles, blocks

35.2.18.   Bricks, tiles and blocks should be stored on firm level bases such as pallets, planking or concrete.

35.2.19.   Piles should not be more than 2.3 m (7 ft 6 in) high.

35.2.20.   Piles higher than 1.2 m (4 ft) high should be stepped back every 30 cm (12 in) above 1.2 m (4 ft).

35.2.21.   When bricks or blocks are removed from a pile they should be so taken that the height of the pile remains uniform and the taper is maintained.

Reinforcing and structural steel

35.2.22.   Reinforcing steel should be stored with only one size in a group so as to facilitate handling.

35.2.23.   Stacks of reinforcing steel should be kept low and adequately blocked to prevent pieces from rolling or the stack from collapsing.

327

Pipes and bars

35.2.24.   Pipes should be stored on racks or in stacks that are blocked so that the pipes cannot roll.

35.2.25.   Large pipes should be removed from the ends and not the sides of stacks.

Earth, sand, gravel, crushed stone

35.2.26.   Piles of loose material such as earth, sand, gravel and rubble should:

(a) slope at the natural angle of repose of the material in question; or

(b) be enclosed by stout fencing.

35.2.27.   The natural angle of repose should if necessary be restored after the addition or removal of material.

35.2.28.   Frozen piles of earth, sand, gravel, etc. should not be undermined.

35.2.29.   Care should be taken that piles of earth, sand, gravel, etc. do not exert dangerous pressure on walls, partitions, etc.

Dusty loose material

35.2.30.   Dusty loose material should be kept in silos, bunkers, bins or the like.

35.2.31.   When dusty loose material is stored or handled, precautions should be taken to prevent the dispersal of dust.

Prefabricated parts

35.2.32.   Prefabricated parts should be stored in such a way as:

(a) to keep them stable; and

(b) to allow them to be easily withdrawn.

36. Working clothes and personal protective equipment

36.1. General provisions

36.1.1.   Where necessary, workers should be provided with and wear protective clothing and other personal protective equipment as conditions may require.

36.1.2.   There should be national standards applicable to personal protective equipment.

36.1.3.   If necessary, workers should be instructed in the use of the personal protective equipment provided.

36.1.4.   Workers should make proper use of and take proper care of the personal protective equipment provided.

36.1.5.   All workers should wear close-fitting clothing and stout boots or other suitable footwear.

36.1.6.   All personal protective equipment should be kept fit for immediate use.

36.1.7.   All necessary measures should be taken by the employer to ensure that protective clothing and the personal protective equipment are effectively worn.

Waterproof clothing

36.1.8.   Workers required to work in the rain or in similar wetting conditions should be provided with waterproof clothing and head covering.

36.1.9.   Oilskin clothing should be kept in a well-ventilated place away from stoves, radiators and other sources of heat, and not be rolled or put in lockers or other confined spaces.

Head protection

36.1.10.   Safety helmets or hard hats should be worn by workers employed at any place where they might be exposed to head injury from:

(a) falling;
(b) falling or flying objects; or
(c) striking against objects or structures.

36.1.11.   Where necessary to prevent danger from electricity, hard hats should be made of insulating material.

36.1.12.   Workers working in the sun in hot weather should wear suitable head covering.

Eye protection

36.1.13.   Workers should be protected by a screen, clear or coloured goggles or other suitable device when employed at places where they might be exposed to eye injury from:

(a) flying particles;
(b) dangerous substances; or
(c) harmful light or other radiation.

Hand and arm protection

36.1.14.   Where necessary, workers should wear suitable gloves or gauntlets or be provided with appropriate barrier creams when employed at places where they might be exposed to hand or arm injuries from:

(a) hot, corrosive or toxic substances; or
(b) sharp or rough points, edges or surfaces of objects.

Foot protection

36.1.15.   Workers should wear footwear of an appropriate type when employed at places where they might be exposed to injury from:

(a) falling or crushing objects;
(b) hot, corrosive or poisonous substances;
(c) sharp-edged tools such as axes;
(d) nails;
(e) abnormal wet; or
(f) slippery or ice-covered surfaces.

Safety belts and lifelines

36.1.16.   Workers who cannot be protected against falls from heights by other means should be protected by safety belts and lifelines.

36.1.17.   Safety belts should be attached by a line of high grade manila or equivalent material to a fixed anchor.

36.1.18.   Where necessary, when working with a safety belt, there should be in addition an independently secured lifeline.

36.1.19.   The lifeline should be anchored above the work to a secure object and the free end should extend to the ground level or working platform.

36.1.20.   Lifelines should be independent of blocks and tackles from which workers may be suspended.

36.1.21.   Only safety belts and lifelines that have been tested in accordance with requirements to be established by the competent authority should be issued to workers.

36.1.22.   All metal parts of safety belts and lifelines should be made of forged steel or equivalent material.

36.1.23.   Safety belts, safety straps, lifelines, permanent anchors and connections should both separately and when assembled:

(a) be capable of supporting safely a suspended load of at least 450 kg (1,000 lb); and

(b) have a breaking strength of at least 1,150 kg (2,500 lb).

36.1.24.   If hooks are used for attaching safety belts to fixed anchors, they should be safety hooks.

36.1.25.   When a lifeline or safety strap is liable to be severed, cut, abraded or burned, it should consist of a wire rope or a wire-cored fibre rope.

36.1.26.   Safety straps should be so fastened to safety belts that they cannot pass through the belt fittings if either end comes loose from its anchorage.

36.1.27.   Metal thimbles should be used for connecting ropes or straps to eyes, rings and snaps.

36.1.28.   Safety belts, safety straps and lifelines should be so fitted as to limit the free fall of the wearer to 1 m (3 ft 3 in).

36.1.29.   Not more than one worker should be attached to any lifeline.

36.1.30.   Safety belts, safety straps and lifelines should be inspected before each occasion of use.

36.1.31.   When a worker's safety depends on a safety belt he should not work in isolation.

Catch nets

36.1.32.   Where workers cannot be protected against falls from heights by other means they should be protected by catch nets.

36.1.33.   Catch nets should be made of good quality fibre cordage, wire or woven fabric or material of equivalent strength and durability.

36.1.34.   The perimeter of catch nets should be reinforced with cloth-covered wire rope, manila rope or equivalent material.

36.1.35.   Catch nets should be provided with adequate means of attachment to anchorages.

Protection against moving vehicles

36.1.36.   Workers who are regularly exposed to danger from moving vehicles should wear:

(a) distinguishing clothing, preferably bright yellow or orange in colour; or

(b) devices of reflecting or otherwise conspicuously visible material.

Protection against drowning

36.1.37.   Life preservers, vests or belts should be worn by workers employed:

(a) on floating pipelines, pontoons, rafts, pilings, coffer cells, stages and the like;

(b) on open-deck floating plant not equipped with bulwarks, guard-rails or other adequate protection;

(c) on structures extending over or adjacent to water and not equipped with guard-rails or other adequate protection;

(d) alone at night at places where they might drown;

(e) in skiffs, small boats and launches if not in the cabin or other enclosed space.

Respiratory protective equipment

36.1.38.   Workers who cannot be protected against airborne dust, fumes, vapours and gases by ventilation or other means should be protected by respiratory equipment.

36.1.39.   Workers employed at places where they might be exposed to injury from lack of oxygen should wear a suitable self-contained air line or respirator.

36.1.40.   All persons required to use respiratory protective equipment should be adequately instructed in its care and use.

36.1.41.   Respiratory protective equipment used by one person should not be used by another before it has been cleaned and sterilised.

36.1.42.   When not in use, respirators should be kept in closed containers.

36.1.43.   Air supplied to airline respirators should be free from harmful contaminants and obnoxious odours.

36.1.44.   When compressed air is used to supply airline respirators:

(a) the compressor should be so placed as to avoid contamination of the air supply;

(b) the air should be supplied at a suitable temperature; and

(c) the compressor should be provided with a safety device to prevent excessive heating so as to eliminate the possibility of generating toxic gases.

36.1.45.   Air supplied to a respirator should not be at an excessive pressure.

36.1.46. In the supply line from a compressor or from a cylinder for compressed air there should be:

(a) a pressure-reducing valve;

(b) a relief valve pre-set to function at a pressure slightly above the setting of the reducing valve if the latter fails; and

(c) a filter that effectively retains pipe scale, oil, water and harmful vapours.

37. Hygiene and welfare

37.1. General provisions

37.1.1.   Shelters, toilet facilities, washing facilities, meal rooms and cloakrooms should:

(a) be adequately lighted and ventilated;

(b) if necessary for reasons of health or welfare, heated; and

(c) maintained in a clean and sanitary condition.

37.1.2.   In all cases, both underground and on the surface, workers should be required to use only the sanitary facilities provided.

37.2. Drinking water

37.2.1.   An adequate supply of cool and wholesome drinking water should be provided for and be readily accessible to all workers.

37.2.2.   All drinking water should be from a source approved by the competent health authority.

37.2.3.   Where such water is not available, the competent health authority should ensure that the necessary steps are taken to make any water to be used for drinking fit for human consumption.

37.2.4.   The use of common drinking cups should be prohibited.

37.2.5.   Drinking water for common use, if stored, should only be stored in closed containers from which the water should be dispensed through taps or cocks.

37.2.6.   Where practicable, hygienic drinking fountains should be provided.

37.2.7.   Supplies of water that is unfit to drink should be conspicuously indicated by notices prohibiting workers from drinking it.

37.2.8.   There should be no means of connecting a supply of drinking water with a supply of water that is unfit to drink.

37.2.9. Water from wells should not be used for drinking unless it has been approved by the competent health authority.

37.2.10. Salt drinks or tablets should be supplied to workers working in great heat, and should be taken as prescribed by a doctor.

37.2.11. If a treatment and purification system is installed to provide drinking water, the system should be approved by the competent health authority before it is used.

37.2.12. If drinking water from an approved public supply has to be transported to the work site, the transport arrangements should be approved by the competent health authority.

37.2.13. The transport and storage tanks and dispensing containers should:

(a) be made of non-corrodible and non-toxic materials, be airtight and be easy to clean;

(b) be cleaned and disinfected at suitable intervals; and

(c) be disinfected in a manner approved by the competent health authority.

37.2.14. Transported and stored drinking water should at all times have a free chlorine residual of not less than one part in a million.

37.2.15. All water dispensed from transport or storage tanks and dispensers should be maintained free from bacteriological contamination.

37.3. Shelters

37.3.1. Suitable shelters should be provided to afford protection for the workers in bad weather.

37.3.2. Shelters should, as far as practicable, provide suitable facilities, unless such facilities are available in the vicinity:

(a) for washing, in conformity with the requirements of section 37.5;

(b) for taking meals, in conformity with the requirements of section 37.6; and

(c) for drying and storing clothing, in conformity with the requirements of section 37.7.

37.3.3. Shelters on the surface should be provided for underground workers.

37.4. Toilet facilities

37.4.1. Adequate toilet facilities should be provided for the workers at easily accessible places.

37.4.2. Toilet facilities should be separate for each sex.

37.4.3. When practicable, water flush toilets connected to public sewage systems should be provided.

37.4.4. No toilet other than a water flush toilet should be installed in any building containing sleeping, eating or other living accommodation.

37.4.5. If a public sewage system is not available, a temporary sewage system should be provided in accordance with the requirements of the competent health authority.

37.4.6. Toilets should be so constructed as to screen the occupants from view and afford protection against the weather and falling objects.

37.4.7. Toilets, including privies, should have a smooth and impervious floor.

37.4.8. For personal cleansing, toilets should be provided with an adequate supply of toilet paper or, where conditions require, water.

37.4.9. Plumbing and other toilet fixtures should comply with the requirements of the competent health authority.

37.4.10. Adequate washing facilities should be provided as near as practicable to toilet facilities.

37.4.11. A sufficient quantity of disinfectants and deodorisers should be provided for chemical closets.

37.4.12. If water flush toilets cannot be provided and privies have to be built:

(a) each privy should be in a fly-tight box over an earth pit with impervious wall;

(b) urinals should discharge directly into the pit through a fly-tight drain.

37.4.13.   No privy should be built within 30 m (100 ft) of any well, or within such greater distance as may be made necessary by the quality of the soil.

37.4.14.   Privies should be disinfected daily.

37.4.15.   The contents of earth-pit privies should be covered daily with sand, lime, wood ash or other suitable material.

37.4.16.   When the contents of an earth pit are within 60 cm (2 ft) of the surface of the ground it should be filled with earth.

37.4.17.   The contents of privy pits should only be removed or buried in compliance with the requirements of the competent health authority.

37.4.18.   Privy pans should have a double flap seat that forms a flyproof joint when closed.

37.4.19.   Chemical closets should comply with the requirements of the competent health authority.

37.5. Washing facilities

37.5.1.   Adequate washing facilities should be provided for all workers, in which:

(a) there should be a sufficient flow of clean water;

(b) there should be adequate means of removing waste water;

(c) suitable non-irritating soap should be supplied in sufficient quantity; and

(d) there should be an adequate supply of drying facilities.

37.5.2.   Washing facilities should not be used for any other purpose.

37.5.3.   Where workers are exposed to skin contamination by poisonous, infectious or irritating substances, or oil, grease or dust, there should be a sufficient number of shower baths supplied with hot and cold water.

37.5.4.   Shower-bath equipment should be thoroughly cleaned at least once in every day of use and effectively disinfected.

37.6. Meal rooms and canteens

37.6.1.   If at least 25 workers are employed on a project, a suitable room in which they can take their own meals should be provided for them unless they can spend mealtimes in their own homes or in another suitable place.

37.6.2.   Meal rooms should be provided with:

(a) a sufficient number of tables and chairs or benches;
(b) drinking water;
(c) adequate facilities for cleaning utensils, table gear, etc;
(d) adequate facilities for heating food and boiling water; and
(e) covered receptacles for the disposal of waste food and litter.

37.6.3.   Receptacles for waste should be emptied after each meal and thoroughly cleaned and, if necessary, disinfected.

37.6.4.   Meal rooms should not be used for any other purpose.

37.6.5.   The floor of meal rooms should be easily washable.

37.6.6.   Dining tables should be covered with suitable non-absorbent washable material and kept clean.

37.6.7.   Meal rooms should be cleaned daily.

37.6.8.   Tables should be cleaned after each meal.

37.6.9.   Where necessary for reasons of health or welfare, a canteen should be provided where workers can obtain hot meals.

37.6.10.   Where necessary, suitable provision should be made to prevent the entry of insects and vermin.

37.7. Cloakrooms

37.7.1.   Cloakrooms should be provided for the workers at easily accessible places.

37.7.2.   Cloakrooms should not be used for any other purpose.

37.7.3.   Cloakrooms should be provided with:

(a) suitable facilities for drying wet clothes;

(b) suitable facilities for hanging clothing including, where necessary to avoid contamination, suitable lockers separating working from street clothes;

(c) benches or other suitable seats.

37.7.4.   Suitable arrangements should be made for disinfecting cloakrooms and lockers in conformity with the requirements of the competent health authority.

37.8. Waste disposal

37.8.1.   A sufficient number of receptacles should be provided at suitable places for the disposal of garbage and other waste.

37.8.2.   Receptacles for waste should be covered, non-corrodible, fly-tight and easy to clean.

37.8.3.   Waste receptacles should be kept closed and emptied at suitable intervals.

37.8.4.   Waste receptacles should be cleaned and disinfected at suitable intervals.

37.8.5.   The contents of waste receptacles should be incinerated, buried or otherwise harmlessly disposed of at suitable intervals.

37.8.6.   Garbage should not be placed or kept elsewhere than in the containers provided.

38. Medical care and supervision

38.1. Medical examination

38.1.1. All workers should undergo as far as practicable a medical examination:

(a) before or shortly after entering employment for the first time (pre-employment examination with special emphasis on physical fitness and personal hygiene); and
(b) periodically, at such intervals which the competent authority should prescribe taking due account of the risks inherent in the work, and the conditions under which the work is performed (periodical re-examination).

38.1.2. All medical examinations should:

(a) be free to the workers; and
(b) include, if necessary, X-ray and laboratory examinations.

38.1.3. Workers under 18 years of age should receive special medical supervision, including regular periodical medical re-examination.

38.1.4. The data obtained by medical examinations should be suitably recorded and kept for reference.

38.1.5. When the work presents a special risk to the health of a worker, he should not be employed on that work.

38.1.6. When a worker is found at the medical examination to constitute a risk to the health or safety of other workers, he should not be allowed to work whilst the risk remains but, if practicable, he should be assigned to work free from such risks.

38.1.7. Workers who have been severely injured or ill should not return to work without permission from a doctor.

38.2. First aid

General provisions

38.2.1. A plan for emergencies and first-aid organisation should be set up in advance for every working area, covering all first-aid personnel and equipment, means of communication, means and ways of transportation.

38.2.2. Every worker should be informed of the plan referred to in paragraph 38.2.1 and instructed accordingly.

38.2.3. Supervisors and/or other responsible personnel should ensure that the plan referred to in paragraph 38.3.1 is strictly adhered to.

38.2.4. First aid in case of accident or sudden illness should be rendered by a physician, a nurse or a person trained in first aid.

38.2.5. Adequate means and personnel for rendering first aid should be readily available at camps, if any, and during working hours at places where work is carried on.

38.2.6. Medical aid should be available on call.

38.2.7. All injuries, however slight, should be reported, treated and recorded as soon as practicable at the nearest first-aid post.

First-aid kits and boxes

38.2.8. First-aid kits or boxes, as appropriate, should be provided at the workplaces and on motor vehicles, locomotives and speeders, and be protected against contamination by dust, moisture, etc.

38.2.9. National regulations or standards should be laid down to specify the minimum contents of first-aid kits and boxes.

38.2.10. Where standards referred to in paragraph 38.2.9 do not exist, the kits or boxes should contain at least compresses and triangular bandages, sterile gauze, antiseptics, adhesive tape, forceps, a tourniquet, blunt-end scissors, splints and, if required, a snake-bite outfit.

38.2.11. First-aid kits and boxes should not contain anything besides material for first aid in emergencies.

38.2.12. First-aid kits and boxes should contain simple and clear instructions to be followed.

38.2.13. First-aid kits and boxes should be in the charge of a responsible person who is qualified to render first aid.

38.2.14. The contents of every first-aid box should be inspected regularly by the person in charge of it, and the box should be kept stocked.

Stretchers or carrying baskets

38.2.15. Stretchers or carrying baskets so constructed that persons can be transported without having to be transferred from the stretcher or the carrying baskets should be readily available.

38.2.16. Two clean blankets should be provided for each stretcher or carrying basket.

Rescue and resuscitation equipment

38.2.17. When workers are employed underground or in other conditions in which they may need to be rescued, suitable rescue equipment should be readily available at or near the construction site.

38.2.18. When rescue equipment may be needed, a sufficient number of trained rescue workers should be constantly available at or near the construction site.

38.2.19. When workers are exposed to risks of drowning or gassing, suitable resuscitation equipment should be kept readily available at or near the construction site.

38.2.20. When resuscitation equipment may be needed, a sufficient number of persons trained to use it should be constantly available at or near the construction site.

First-aid rooms

38.2.21. If as a rule 100 or more workers are employed on any shift at least one suitably equipped first-aid room or station should be provided at a readily accessible place for the treatment of minor injuries and as a rest place for seriously sick or injured workers.

38.2.22. A responsible person qualified to render first aid should be in charge of the first-aid room or station and be readily available during working hours.

Ambulances

38.2.23. Arrangements should be made to ensure the prompt transport, where necessary, of sick or injured workers to a hospital or other equivalent treatment centre.

38.2.24. Where practicable, such arrangements should include facilities for promptly obtaining an ambulance carriage from some place situated within a reasonable distance of the working area.

38.2.25. If an ambulance is not available other reasonably comfortable means of transport should be provided.

Notices

38.2.26. Notices should be conspicuously exhibited at suitable places stating:

(a) the position of the nearest first-aid box, first-aid room, ambulance and stretcher and the place where the person in charge can be found;

(b) the place of the nearest telephone for calling the ambulance, and the telephone number and name of the person or centre to be called; and

(c) the name, address and telephone number of the doctor, hospital and rescue station to be called in an emergency.

First-aid personnel

38.2.27. All supervisors should receive instruction in first aid.

38.2.28. Workers should be encouraged to take first-aid training when available.

Register

38.2.29. A first-aid register should be kept in each first-aid room for recording the names of persons to whom first aid has been rendered and the particulars of injuries and treatment.

38.2.30. The first-aid register should only be accessible to authorised persons.

38.3. Medical services

38.3.1. If the workers live in a camp, the employer should be responsible for emergency medical care, including the provision of medicines and removal to hospital.

38.3.2. The employer may arrange for the necessary services to be provided by medical practitioners and hospitals.

38.3.3. Employers should provide for:

(a) first-aid and emergency treatment;

(b) pre-employment, periodical and special medical examinations;

(c) periodical training of first-aid personnel;

(d) surveillance of, and advice on, all conditions at workplaces and facilities that affect the health of workers; and

(e) promotion of health education among workers.

38.3.4. Medical services should be directed by a doctor and should be provided with an adequate staff of appropriate para-medical personnel.

38.3.5. Nurses employed in the medical service should possess a certificate of proficiency acceptable to the competent authority.

38.3.6. The premises occupied by the medical service should:

(a) be at ground level;

(b) be conveniently accessible from all workplaces;

(c) be so designed as to allow stretcher cases to be handled easily; and

(d) so far as practicable not be exposed to excessive noise.

38.3.7. The premises should comprise at least a waiting room, a treatment room, a rest room and toilet and washing facilities.

38.3.8.   The rooms referred to in paragraph 38.3.7 should:
(a) be sufficiently spacious, suitably lit and ventilated, provided
     with drinking water and, where necessary, heated or cooled;
(b) have washable walls, floor and fixtures.

38.3.9.   Rest rooms should be provided with beds and be so
located and arranged as to allow their use for isolation purposes.

38.3.10.   The medical service should keep such records of its
activities as will provide adequate information on:
(a) the workers' state of health; and
(b) the nature, circumstances and outcome of occupational
     injuries.

39. Construction camps

39.1. General provisions

39.1.1. On construction sites employing considerable numbers of workers at places remote from their homes or other suitable living accommodation, employers should provide suitable accommodation.

39.1.2. The competent authority should be notified of the opening of any construction camp.

39.1.3. Construction camps should be maintained in a good state of repair, and in a clean and sanitary condition.

39.1.4. The employer should appoint a competent person to be in charge of the camp and to be responsible for its proper maintenance.

39.1.5. Camp sites should:

(a) be properly drained;

(b) be cleared of trees that might cause danger by falling;

(c) be at a sufficient distance from animal pens, stables, sheds, accumulations of refuse, manure or other offensive matter;

(d) be at a safe distance from construction railways and roads; and

(e) be at an adequate distance from streams, lakes, springs, wells and other watercourses, and from potential sources of water pollution.

39.1.6. Living accommodation should be sufficient and suitable, and, in particular:

(a) the accommodation should be effectively protected from the weather, ground moisture and vermin;

(b) the sleeping quarters should be separate from the dining quarters;

(c) there should be a sufficient supply of furniture and the necessary utensils;

(d) suitable provision should be made for supplies of drinking water and washing water;

(e) suitable provision should be made for lighting, ventilation, sanitation and, if necessary, heating;

(f) suitable provision should be made for storing perishable provisions;

(g) suitable provision should be made for washing and drying clothes;

(h) suitable first-aid material should be provided;

(i) suitable provision should be made for the hygienic disposal of kitchen garbage, and for drainage from the dining, cooking and washing quarters and toilet facilities; and

(j) camps should be constructed and maintained in conformity with regulations to be laid down by the competent authority concerning protection against fire.

39.1.7. The supply of drinking water should comply with the requirements of section 37.2.

39.1.8. Toilet facilities should comply with the requirements of section 37.4.

39.1.9. Washing facilities should comply with the relevant requirements of section 37.5.

39.1.10. Meal rooms should comply with the requirements of section 37.6.

39.1.11. Arrangements for waste disposal should comply with the requirements of section 37.8.

39.1.12. Heating installations should comply with the requirements of paragraphs 2.4.13 to 2.4.23.

39.1.13. In living quarters the use of braziers should be prohibited.

39.1.14. Camps should be provided with a recreation room.

39.1.15. Workers should take good care of the living accommodation and its equipment and should not wilfully damage or dirty it.

39.1.16.   No building, structure or enclosure in a construction camp should be used for the manufacture, storage or handling of toxic and other harmful substances.

39.1.17.   Camps should be inspected at sufficiently frequent intervals for vermin.

39.1.18.   Premises or persons found to be infested with vermin should be suitably treated.

39.1.19.   As soon as any communicable disease occurs or is suspected in a camp the employer should notify the competent health authority.

39.1.20.   Caves, straw huts, tents—unless specifically designed for the purpose—storerooms and stables should not be used as living accommodation.

39.1.21.   Boats used as living quarters should comply with the relevant provisions of this chapter.

39.2. Sleeping quarters

39.2.1.   Adequate and properly ventilated sleeping quarters should be provided.

39.2.2.   Where necessary, sleeping quarters should be protected against the penetration of animals, mosquitoes and other winged insects.

39.2.3.   A separate bed should be provided for each worker.

39.2.4.   Beds should:

(a) be suitably elevated from the floor; and
(b) be provided with a mattress or sleeping bag, a pillow and the necessary sheets and blankets.

39.2.5.   Bedding should be kept in good condition and washed at suitable intervals.

39.2.6.   The bedding should be washed and disinfected:

(a) when occupancy of a bed changes; and
(b) when an occupant contracts an infectious or contagious disease.

39.2.7.   Walls of sleeping quarters should be easily washable.

39.2.8.   Floors of sleeping quarters should:

(a) be of impermeable material; and
(b) be adequately raised from the ground.

39.2.9.   Sleeping quarters should be cleaned daily by a method that does not raise dust, for instance a vacuum method or a damp method.

39.2.10.   Sleeping quarters and bedding should be disinfected at suitable intervals.

39.2.11.   Main meals should not be eaten in sleeping quarters.

39.3. Catering

39.3.1.   Furniture, equipment and appliances of kitchens and dining rooms should be so constructed and installed as to facilitate thorough cleaning, and the maintenance of the kitchen or dining room in a clean and sanitary condition.

39.3.2.   Kitchens and dining rooms should not be used for any purpose other than preparing, storing, serving or consuming food.

39.3.3.   All food should be adequately protected against contamination and deterioration.

39.3.4.   Utensils and tableware used in the preparation, serving, storage or consumption of food should be adequately cleaned after each occasion of use.

39.3.5.   Dining rooms and kitchens should be adequately ventilated.

39.3.6.   Dining rooms and kitchens should be constantly maintained in a clean and sanitary condition.

39.3.7.   Cooks and other persons handling or preparing food should be free from any communicable disease.

39.4. Medical facilities

39.4.1. In camps, where necessary there should be suitable accommodation for sick persons.

39.4.2. Hospitals or infirmaries should comply with the requirements of section 38.3.

40. Safety organisation

40.1. General provisions

40.1.1. On all projects on which 25 or more workers are regularly employed, the employer should appoint a safety officer to be in charge of all matters relating to safety and hygiene on the project.

40.1.2. On all projects on which 250 or more workers are regularly employed, the safety officer should be employed full time on safety and health activities.

40.1.3. Safety committees should be established in any project where the circumstances warrant it.

40.1.4. Safety officers should compile an individual report containing full information on the causes and circumstances of every lost-time accident, minor accident and dangerous occurrence with a view to preventing recurrences.

40.1.5. Copies of the report referred to in paragraph 40.1.4 should be sent to the management.

40.1.6. Safety officers or safety committees should:

(a) consider circumstances and causes of all accidents occurring on a project;

(b) make recommendations to the employer for preventing the occurrence or recurrence of accidents;

(c) make periodical inspections of the work site and all its equipment in the interests of safety and hygiene;

(d) watch over the execution of particular measures taken for the prevention of accidents;

(e) watch over compliance with official regulations, instructions, etc. relating to safety and hygiene;

(f) endeavour to secure the co-operation of all workers in the promotion of safety and hygiene;

(g) participate in the drawing-up of the undertaking's safety rules;

(h) study the statistics of accidents occurring on the project;
(i) see that all new workers, and workers transferred to new jobs, receive adequate safety training, instruction and guidance; and
(j) if necessary to prevent danger, report to the competent official inspector any unsatisfactory conditions as regards safety and health that the employer fails to remedy within a reasonable time.

40.1.7. Safety committees should consist of representatives of the employer and of the workers and should include:

(a) a high executive official;
(b) the safety official or officials;
(c) foremen; and
(d) a representative of the undertaking's medical service if there is one.

40.1.8. The workers' representatives on safety committees should be elected by all the workers in such a manner that all suitably qualified workers are enabled to serve on a committee in turn.

40.1.9. Safety committees should meet at suitable intervals and keep adequate records of all meetings.

40.1.10. Employers should:

(a) give safety committees all reasonable encouragement and facilities in the performance of their duties;
(b) consult safety committees in all matters relating to safety and health on the project;
(c) take all practicable steps to give effect to recommendations of the safety committee; and
(d) in cases where they do not adopt a recommendation of the safety committee, inform the committee of the reasons within a reasonable time.

40.1.11. On all projects records should be kept of all lost-time accidents, minor accidents and dangerous occurrences.

40.1.12. The records should include statistics that will:

(a) show the accident record of each operation, occupation and individual; and

(b) show the distribution of accidents by causes.

40.1.13. Accident statistics should be compiled by methods approved by the competent authority so as to ensure their comparability with those for other projects and construction undertakings.

40.1.14. Where appropriate, employers should make arrangements whereby workers can submit suggestions relating to safety and health on the project.

40.1.15. Where two or more employers are engaged on a project they should co-ordinate their safety activities by:

(a) the appointment of a joint safety officer;

(b) the appointment of a joint safety committee; or

(c) other effective means.

40.1.16. Employers' and workers' organisations should, in the course of their joint activities, pay particular attention to the question of safety and health propaganda, and should endeavour to improve all safety measures by this means.

41. Miscellaneous provisions

41.1. Workshops

41.1.1.   Maintenance and repair shops and other workshops should comply with:

(a) national or other official regulations concerning occupational safety and health in industrial establishments; or

(b) in so far as concerns matters not dealt with in such regulations, with the *Model Code of Safety Regulations for Industrial Establishments* published by the International Labour Office.

41.2. Land clearance

41.2.1.   Land-clearance operations should comply with the relevant requirements of the Code of Practice *Safety and Health in Forestry Work* published by the International Labour Office.

41.3. Poisonous plants, insects, snakes, etc.

General provisions

41.3.1.   In regions infested with poisonous plants, dangerous insects or venomous snakes, workers should be taught how to identify them, and instructed in precautionary measures, symptoms of illness and emergency first-aid treatment.

Poisonous plants

41.3.2.   Persons known to be hypersensitive to plant poisons should not be employed in regions infested with poison oak, poison ivy, poison sumac or other poisonous plants.

41.3.3.   Persons working in regions infested with poisonous plants should keep as much of the body covered as practicable, as by wearing tight-fitting clothing, gauntlets and leggings.

41.3.4.   After the day's work:

(a) exposed parts of the body should be thoroughly washed with soap and water;

(b) clothes should be dry cleaned or washed; and

(c) tools should be cleaned.

41.3.5.   Poisonous plants around camps and other places where workers congregate should as far as practicable be destroyed.

41.3.6.   If poisonous plants are burned:

(a) an isolated place should be chosen for the purpose; and

(b) workers should avoid all contact with the smoke and residual ash.

Insects, etc.

41.3.7.   Persons working in regions infested with dangerous insects should keep as much of the body covered as practicable, as by wearing tight-fitting clothing, gauntlets and leggings.

41.3.8.   In regions infested with ticks, workers should:

(a) inspect the body and clothing at least once a day;

(b) at night ensure that ticks cannot get into clothing or beds; and

(c) get medical treatment if they become feverish.

41.3.9.   Ticks found on the body should be removed, if possible without causing any skin puncture.

41.3.10.   In regions infested with chigoes (chiggers), workers should:

(a) avoid low vegetation if practicable;

(b) avoid sitting on the ground or on logs;

(c) dust legs and arms with sulphur and take sulphur tablets;

(d) use insect repellents such as dimethyl phthalate;

(e) take a hot bath every day; and

(f) get medical treatment immediately if a bite becomes rapidly inflamed.

41.3.11.   In regions infested with black-widow or other poisonous spiders, workers should:

(a) wear gloves;

(b) inspect objects before handling them; and

(c) inspect outdoor toilets before using them.

Snakes

41.3.12.   In regions infested with venomous snakes, workers should:

(a) always carry a snake bite kit;

(b) wear high boots;

(c) keep a good look-out in surroundings where snakes could be concealed by foliage, rocks, logs, etc.;

(d) move stacked lumber and other stacked material with an appliance such as a bar, and not with the hands; and

(e) if bitten keep still and use the kit as directed.

41.4. Intoxicants

41.4.1.   Persons under the influence of alcohol or other intoxicants should not be allowed to work in construction operations.

41.4.2.   No alcohol or other intoxicants should be furnished to workers while they are engaged in construction operations.

41.5. Reporting and investigation of occupational accidents and diseases

41.5.1.   All accidents to workers causing loss of life or serious injury should be reported forthwith to the competent authority.

41.5.2.   Other injuries and occupational diseases causing incapacity for work should be reported to the competent authority within such time and in such form as may be specified in national or other official regulations.

41.5.3.   Dangerous occurrences such as explosions, collapse of cranes or derricks and serious fires as may be specified in national or other official regulations should be reported forthwith to the competent authority whether any personal injury has been caused or not.

41.5.4.   When a fatal accident has occurred, the scene of the accident should as far as practicable be left undisturbed until it has been visited by a representative of the competent authority.

41.5.5.   Plant or gear on which a dangerous failure has occurred should as far as practicable be kept available for inspection by the competent authority.

INDEX

earth-moving equipment
15.1.23-15.1.24
electrical installations   17.7
equipment, general requirements
2.9.6-2.9.13
hand tools   16.1.5-16.1.7
hoists   5.2.40-5.2.41
ladders   4.1.12-4.1.15
lifting appliances   5.1.26-5.1.28
machines   13.2
motor vehicles   10.5
near machines   13.2.7
pile drivers   24.2
power shovels, excavators
15.2.23-15.2.24
railways   9.1.25-9.1.28
scaffolds   3.1.23-3.1.27
silos   20.2.7
structures, general requirements
2.9.6-2.9.13
tractors   10.5
Man locks
see Locks, man
Manufacturers, obligations of   1.5
Marking, for containers of dangerous substances   21.1.9
Masks
see Protective equipment, personal, respiratory
Masts, concrete bucket   25.4.1;
25.4.25-25.4.26
Materials
carrying, manual   35.1
combustible   2.4.24-2.4.28; 21.2
dangerous, general provisions   21
handling, general provisions   35
for hand tools   16.1.1-16.1.4
lifting, manual   35.1
piling   35.2
for scaffolds   3.1.3-3.1.11
stacking   35.2
Maximum safe load
see Load, maximum safe
Meal rooms
see Dining rooms

Medical examination
see Examination, medical
Medical locks
see Locks, medical
Medical services
general requirements   38.3
for camps   39.4
Medical supervision
see Supervision, medical
Metal parts of structures   2.9.3
Methane   32.1.9
Mineral wool   29.10.11-29.10.12
Misfires
blasting   23.7.5-23.7.6
powder-actuated tools   16.3.40-
16.3.41
Mixers
asphalt   15.5.4-15.5.7
concrete   15.8
Mobile scaffolds
see Scaffolds, mobile
Monorail hoists
see Hoists, monorail
Motors, electric   17.4.19-17.4.22
Motor trucks
see Trucks, motor

Nails
projecting in wood, etc.   2.3.2
on scaffolds   3.1.8; 3.1.21-3.1.22
Nets, catch   36.1.32-36.1.35
Noise   2.7
Notices
general requirements   1.2.8-1.2.9
for dangerous atmospheres
21.3.9
for compressors   18.2.1; 33.1.5
for diving stations   34.1.13
for earth-moving equipment
15.1.1
for electrical installations
17.2.26-17.2.27
first aid   38.2.26
for hoists   5.2.39